The

Quranic

narrative on

Human Evolution

Divine Exploration of Human Evolution

Saeed Rajput

Copyright © 2024 Saeed Rajput - All rights reserved.
ISBN: 978-1-3999-8255-9

The content within this book is protected by copyright law. Individuals are permitted to reproduce and distribute the material contained herein solely for non-commercial purposes. Any unauthorized use, including modification or commercial exploitation, is prohibited without explicit authorization from the author.

Dedicated to

My love for the Quran and Science

Table of Contents

Preface 8
PART I: Introduction 11
Chapter One: Quran 11
    History 11
    Objective 13
    Literature 13
    Scientific Parallels in the Quran 16
Chapter Two: Science of Human Evolution 21
    Theories and Concepts 21
    Taxonomy of Human Evolution 37
    Stages of Human Evolution 39
    Fossil Discoveries 42
    Contribution of Muslim Scientists 49
Chapter Three: Science vs Religion 53
PART II: Research 56
Chapter Four: Extraction of Verses 59
    Group 1: Origin of Life 59
    Group 2: Creation with Mud 63
    Group 3: Creation with Water 68
    Group 4: Stages of Creation 69
    Group 5: Bashar (Pre-human being) 72

Group 6: First Human Being 74
Group 7: Adam 78

Chapter Five: Origin 95
Nafs-in Wahid (A Single Soul) 95
Creation from Clay 99
Creation with Water 100

Chapter Six: Bashar 106
Human Descent from Apes? 108

Chapter Seven: First Human Being 111

Chapter Eight: Adam 120
Story of Adam in the Quran 120
Adam – Father of Humanity 122
Parents or No parents? 129
Place of Creation and Test of Adam 131
Others Coexisted with Adam and Eve? ... 137
Era of Adam 141
Reconstructed Story of Adam 143

Chapter Nine: Summary of Research 145
Origin of Life 145
Reproduction 147
Bashar or Homo sapiens 148
Human being or Homo sapiens sapiens ... 149

Adam – The First Human being 150
PART III: Conclusion 154
    Unbiased Pursuit of Knowledge 154
    Illuminating Quranic Revelations 154
    The Constant of the "Will of God" 156
    Science and Interpretation of the Quran .. 157
    Limitations of Science 158
    Impact of new Scientific discoveries 159
    Lack of Science Education 160
    Blind Faith and Polarization 161
    Persecution and Adversity 162
    Healthy Debates 163
About the Author 165
Bibliography 166

Preface

In the quest for insight regarding the Quran's perspective on the evolution of human beings, an inherent drive compelled me to embark on an independent journey of discovery. The reservoir of available knowledge appeared inadequate, the scholars had merely grazed the surface of this deep topic, leaving behind fragments of divergent viewpoints. This array of conflicting viewpoints not only created more confusion but also fueled an even stronger desire to fully grasp the depths of the subject.

In adherence to tradition, as a non-Arabic-speaking Muslim, I began learning to read the Quran at a very young age. However, the focus was solely on reading the words, without delving into the profound meanings concealed within the Quranic verses. Such an approach remains prevalent in the households of non-Arabic-speaking Muslims to this day. It wasn't until my teenage years that a newfound curiosity arose within me—to truly grasp the message within. I distinctly recall an instance when I listened to a speech by Dr. Zakir Naik, a widely acclaimed Muslim scholar, on television. It was a debate on the topic of the connection between the Quran and Science, a conversation that ignited a spark of fascination within me for a deeper exploration of the Quran. It was then that I made the decision to seek the Quran's message through its translation.

However, relying solely on translation proved insufficient for genuine understanding until I encountered the enlightening lectures by Dr. Asrar Ahmed, another respectable scholar of Islam. His clear explanations of Quranic verses made the intricate text accessible. I found myself captivated by the profound depths of its literary richness, the artful use of metaphorical expressions, and the divine wisdom woven throughout. Since then, I've revisited the Quran multiple times, and each reading has been a fresh journey. Every time, I've uncovered something new—an enduring testament to the Quran's inherent beauty.

I regard Dr. Asrar Ahmed as both my mentor and guide, someone who encouraged me to open my mind to broader horizons of thought. It was his lectures that urged me to consider the idea of evolution in human creation for the first time. Until then, my understanding of human creation had been confined to a rejection of any notion of evolution. Through his explanation of Chapter 3 Verse 33 of the Quran, *"Truly God chose Adam and Noah and the descendants of Abraham and of Imran above all mankind,"* he proposed an interesting thought. He suggested that choosing Adam could accommodate the viewpoint of those who speculate that preceding Adam, there may have existed other species. This was the catalyst that initiated my quest for further literature on this subject. Unfortunately, there wasn't much information available but, I resolved to undertake a personal exploration of the Quran's teachings, seeking divine insights into the matter of human evolution.

The purpose of this book is to delve into the wisdom offered by the Quran over fourteen centuries ago, contributing to the ongoing discourse and various theories surrounding the origins of humankind. It aims to achieve a profound understanding of our ancestors and to grasp the stages of survival they traversed—a journey that has significantly influenced the shaping of our present.

The Quran employs metaphors of profound significance, yet they require careful handling to prevent misinterpretations. Precautionary measures are taken to navigate away from any potential misunderstandings. The text draws parallels between the metaphorical explanations in the Quran and the esteemed theories put forth by renowned scientists in this field. This synthesis also encompasses the contemporary spectrum of discoveries, aligning the wisdom of the Quran with the frontiers of modern knowledge.

In general, theology and the science of human evolution are often perceived as fundamentally divergent fields. However, an effort is made to critically examine this perceived dichotomy, striving to determine whether this perception is accurate. Throughout this exploration, meticulous care is exercised to set aside personal preferences and biases, adopting a neutral stance in the quest for an objective understanding.

PART I: Introduction

An introduction is essential before delving into the heart of the subject matter. Before embarking on an in-depth exploration, it is imperative to provide a foundational understanding of the Quran and offer a preliminary overview of the science of Human Evolution, encompassing both prevailing theories and established empirical facts.

Chapter One: Quran

At its core, the Quran transcends the status of a mere book. For over 2 billion Muslims, it carries the profound weight of being the literal words of God. It serves as a cornerstone, a catalyst in the meteoric ascent of Islam, propelling it to become the world's fastest-growing religion. Asserting its claim of remaining unchanged since its inception centuries ago, the Quran goes beyond being a static book. It functions as a foundational constitution, skillfully shaping legal paradigms. Moreover, it stands as a way of life, steadfastly embraced by Muslims in every corner of the world.

History

In the 7th century Arabia, in the city of Mecca, an individual named Muhammad, a forty years old illiterate man, renowned in his community as Sadiq – Truthful and Ameen – Trustworthy, claimed to receive divine revelations from God through an angel named Gabriel. These divine messages spanned over 23 years and were eventually compiled into a sacred book known as the Quran by his devoted followers.

The revelations within the Quran defined a clear distinction between two epochs: the era preceding its arrival and the era that followed. The pre-Quranic epoch was characterized as a time shrouded in darkness and moral decay. However, with the advent of the Quran, a profound transformation swept across the Arab world, toppling the prevailing order. This transformation then radiated outward, impacting even the mighty superpowers of the time, Rome and Persia.

In the era of pre-Islamic Arabia, the Arabs were characterized by a multitude of moral transgressions. It was a land with no centralized state or a single governing body, instead a tribal system prevailed, marked by never-ending conflicts and long-standing enmities that transcended generations. Slavery was not only common but also hideously inhumane, as slaves were regarded as mere chattel, subject to the absolute authority of their masters, who wielded the power to dictate every aspect of their lives, including the grim ability to take their lives without facing any consequences.

Women endured marginalization, devoid of rights and respect, while their lives were tightly regulated. Men, in stark contrast, enjoyed the privilege of multiple marriages and significant discretion in their treatment of wives. Women had no entitlement to inheritance, and the birth of female infants was often viewed as an unwelcome burden, tragically resulting in the abhorrent practice of infanticide, including the cruel act of burying them alive. The legal system exhibited partiality, favoring the wealthy and influential. In sum, this society was marked by a profound absence of ethics and morality.

Amidst this era shrouded in darkness, the Quran emerged as a profound beacon of hope, leaving its profound impression on the most vulnerable and marginalized members of society first. What followed was a sweeping challenge to the existing societal structure, a determined dismantling of its entrenched and abhorrent practices one by one. In a remarkably short span, this society underwent a profound transformation, making a complete 180-degree turn. It evolved into a paragon of justice, ethics, and moral rectitude, extending rights not only to women and slaves but even to the welfare of animals. The Quran proclaims a sense of brotherhood and sisterhood among individuals, thereby transforming generations-long enmities into bonds of fraternity.

Objective

The Quran's sole objective is Hidayah, which translates to guidance or enlightenment in Arabic. This guidance extends to all aspects of human life, including the spiritual, moral, ethical, and even practical dimensions. It aims to lead individuals toward a life of righteousness, compassion, and a profound connection with the Divine.

To accomplish this paramount goal, it offers explicit directives on what to do and what to avoid. These directives encompass issues such as honesty, justice, kindness, and compassion, while also prohibiting actions like theft, murder, and deceit. In this way, the Quran sets a moral compass for individuals and societies, aiming to create a harmonious and just world.

It draws upon the lessons of history, recounting the fates of past civilizations that met their downfall as a punishment of their wrongdoings. Furthermore, it imparts the promise of a brighter afterlife to those who adhere faithfully to the path charted within its verses.

A remarkable feature of the Quran's guidance is its emphasis on the unity of humanity. It fosters a sense of universal brotherhood. This teaching aims to break down barriers of race, ethnicity, and nationality, encouraging compassion and solidarity among all people.

The Quran's objective is clear and profound. It is a beacon of light in a complex world, offering individuals and societies the tools to navigate life's challenges while striving for spiritual fulfilment and righteousness. The Quran stands as a timeless and transcendent source of wisdom, impacting the lives of millions and guiding humanity toward a more just, compassionate, and harmonious existence.

Literature

The Quran not only serves as a spiritual guide but also stands as an unparalleled literary masterpiece, a testament to profound artistry.

Certain individuals in Mecca initially posited the idea that the Quran could be construed as poetry; however, the Quran decisively rebutted this notion. In Chapter 36 Verse 69, it is explicitly stated, *"And We did not give Prophet Muhammad knowledge of poetry, nor is it befitting for him. It is none but an admonition and a clear book."* This clear declaration dispels any notion that the Quran is a work of poetry, even though certain verses exhibit rhyming patterns.

So, is the Quran's style akin to prose? The Quran is structured in such a distinctive manner that it doesn't merely progress linearly from one point to the next in a sequence. At times, the selection of topics within a chapter appears to be arranged in a manner that might seem random to the uninitiated reader. For instance, a verse may address a wrongful deed, followed by one related to the afterlife, and then the next entirely separate topic. Thus, it becomes evident that the Quran's style is not that of prose either.

The literary style of the Quran can be aptly described as oration. Each chapter represents a distinct oration, and when taken together, the Quran can be viewed as a compilation of divine speeches. This characteristic imbues the Quran with a truly unique style as a book. It closely resembles a sermon, where the speaker approaches with a clear objective: to impart a specific message to the audience. Throughout the oration, themes are intricately woven together, emotions are stirred, and the audience is simultaneously uplifted and prompted to engage in thoughtful reflection. The speaker motivates, warns, and encourages deep thought. By the end of the sermon, the speaker achieves the intended objective, just as in the Quran, where each divine message is delivered with precision and purpose.

Written in Arabic, it possesses a linguistic richness that is truly awe-inspiring. Each word, phrase, and verse is meticulously chosen to convey the intended message with utmost clarity and depth. This precision is not only appreciated for its linguistic beauty but also for its ability to convey profound spiritual and ethical insights. The Quran has bestowed a remarkable favor upon the Arabic language, preserving it in the same pristine form as it existed, fourteen centuries ago, at the time of the Quran's revelation.

Another distinctive quality of the Quran is its inimitability–the impossibility of being replicated or equaled by human endeavor. The Quran boldly issues this challenge in Chapter 17 Verse 88, *"Were all mankind to come together and wish to produce the like of the Qur'an, they would never succeed, however much they aided each other."* Subsequently, this challenge was modified to require the creation of ten chapters of similar calibre in Chapter 11 Verse 13, *"Do they say: He (Muhammad) has invented this Book himself? Say: If that is so, bring ten surahs (chapters) the like of it of your composition, and call upon all (the deities) you can other than God to your help. Do so if you are truthful."*

Eventually, the challenge was narrowed down to replicate just one chapter in Chapter 2 Verse 23 of the Quran, *"If you are in any doubt whether it is We Who have revealed this Book to Our servant, then produce just a surah (chapter) like it, and call all your supporters and seek in it the support of all others but God. Accomplish this if you are truthful,"* and Chapter 10 Verse 38 of the Quran states, *"Do they say that the Messenger has himself composed the Qur'an? Say: In that case bring forth just one surah like it and call on all whom you can, except God, to help you if you are truthful."* Yet, this challenge has remained unmet throughout history. This singular feature of the Quran reinforces its divine origin, as it is widely recognized as beyond human capacity to replicate its literary excellence.

Another noteworthy feature of the Quran is designed to be comprehensible and memorable, as affirmed in Chapter 54 Verse 17, *"And We have certainly made the Quran easy for remembrance, so is there any who will remember?"* This ease of memorization and understanding has been demonstrated by millions of Muslims who have committed the entire Quran to memory. This achievement underscores the Quran's broad appeal and its unique capacity to be readily embraced and retained by individuals, setting it apart from other literary works.

Scientific Parallels in the Quran

The Quran is not primarily a book of science. However, it is interesting to note that within its pages, there exist numerous references that seem to harmonize with modern scientific knowledge. This intriguing alignment between the Quranic text and modern scientific understanding raises questions and curiosity.

Considering the Quran's claim to be the very words of God, it carries an inherent expectation that it should not conflict with established scientific facts. In this light, the Quran stands as a unique religious text that invites exploration and observation from both religious and scientific perspectives.

While a wealth of literature explores the intersection of the Quran and science, it's important to clarify that this introduction aims to provide but a glimpse of this vast topic. It will do so by selectively highlighting a few compelling examples from the Quran that appear to resonate with contemporary scientific understanding. These examples are intended to serve as a starting point for further exploration, encouraging readers to delve deeper into the fascinating interplay between the Quran and modern science.

Water:

Chapter 21 Verse 30 states, *"We made every living thing from water; will they not believe?"*

Following the invention of the microscope, scientific inquiry discovered that the fundamental composition of living organisms predominantly comprises water. Within the cell, the cytoplasm consists of a remarkable 80% water content. Subsequent modern research has further illuminated this aspect, demonstrating that most organisms encompass a striking range of 50% to 90% water within their structure. This revelation underscores a critical aspect of life itself, emphasizing that water is an indispensable necessity for the existence of every living entity.

Big Bang Theory:

Chapter 21 Verse 30 states, *"Have those who disbelieved not considered that the heavens and the earth were a joined entity, and We separated them."*

In the year 1965, distinguished radio astronomers Arno Penzias and Robert Wilson were honored with the Nobel Prize for their groundbreaking discovery that provided substantial confirmation for the Big Bang theory, explaining the cosmic origins of the universe. This profound theory suggests that the universe originated from an exceedingly compact singularity, a single point, before undergoing a dramatic expansion and stretching, ultimately reaching the vast expanse we perceive today.

Embryology:

Chapter 23 Verses 12-14 state, *"We created man from an extract of clay. Then We made him as a drop in a place of settlement, firmly fixed. Then We made the drop into an alaqah (leech, suspended thing, and blood clot), then We made the alaqah into a mudghah (chewed substance) ..."*

Professor Emeritus Keith L. Moore, a globally renowned scientist in the domains of anatomy and embryology, articulated a perspective that underscores the remarkable nature of the Quranic revelations. He asserted, "It is clear to me that these statements must have come to Muhammad from God because almost all of this knowledge was not discovered until many centuries later." Professor Moore's acknowledgement stands as a testament to the Quran's ability to impart knowledge that predates the scientific discoveries of many centuries.

Iron:

Chapter 57 Verse 25 states, *"We sent down Iron with its great inherent strength and its many benefits for humankind."*

As articulated by M. E. Walrath, the presence of iron on Earth is not a natural occurrence. Scientific consensus asserts that billions of years in the past, our planet experienced impacts from meteorites. Within these cosmic wanderers, iron was originally found. The collisions of these meteorites with Earth have granted us access to this essential element.

Meeting of the Seas:

Chapter 55 Verses 19-20 state, *"He released the two seas, meeting [side by side], Between them is a barrier [so] neither of them transgresses."*

Through scientific exploration, it has come to light that at the juncture where two distinct bodies of water converge, a natural barrier exists. This barrier plays a crucial role in preserving the unique characteristics of each sea, including their respective temperature, salinity, and density. It serves as a natural demarcation, ensuring that these distinct marine environments maintain their individual properties.

Heavenly Orbits:

Chapter 21 Verse 33 states, *"And it is He who created the night and the day and the sun and the moon; all [heavenly bodies] in an orbit are swimming."*

While the concept was primarily embraced by astronomers in the 20th century, it has now evolved into a firmly established scientific consensus that the Sun, the Moon, and all celestial bodies within the Universe are indeed engaged in perpetual orbital motion. This understanding marks a significant departure from the previously held notion of celestial bodies as stationary entities, underscoring the dynamic nature of the cosmos.

Mountains as Stakes:

Chapter 78 Verses 6-7 state, *"Have We not made the earth a resting place? And the mountains as stakes?"*

In the book "Earth" authored by geophysicist Frank Press in 1986, an intriguing analogy is presented regarding mountains. It likens these natural formations to stakes, emphasizing that their substantial portions are concealed beneath the Earth's surface. For instance, Mount Everest, towering at an approximate elevation of 9 kilometers above sea level, possesses an extensive subterranean root that extends to depths exceeding 125 kilometers.

Expansion of the Universe:

Chapter 51 Verse 47 states, *"And the heaven We constructed with strength, and indeed, We are [its] expander."*

In the year 1922, Alexander Friedmann harnessed Einstein's field equations to furnish compelling theoretical proof of the universe's expansion. Subsequently, in 1924, Swedish astronomer Knut Lundmark achieved the distinction of being the first to distinguish observational evidence supporting this cosmic expansion.

Pain Receptors:

Chapter 4 Verse 56 states, *"We shall send those who reject our revelations to the (hell) fire. When their skins have been burned away, We shall replace them with new ones so that they may continue to feel the pain: God is almighty, all-wise."*

For a long time, the prevailing belief was that our ability to feel sensations and pain was contingent upon the brain's function. Yet, recent discoveries have unveiled the existence of pain receptors within the skin. These pain receptors play a pivotal role, as they enable an individual to experience the sensation of pain.

Finger Prints:

Chapter 75 Verses 3-4 state, *"Does man think that We cannot assemble his bones? Nay, We are able to put together in perfect order the very tips of his fingers."*

In the year 1788, Johann Christoph Andreas Mayer, a distinguished German anatomist, became the first European scholar to assert the distinctive uniqueness of fingerprints for each individual. Then, in 1880, Henry Faulds, through his extensive research, proposed that fingerprints indeed possess a unique characteristic exclusive to humans.

Chapter Two: Science of Human Evolution

In the field of biology, evolution refers to the process marked by the modification of inheritable characteristics within biological populations as they progress through successive generations.

The story of human evolution is a captivating narrative that spans countless millennia, revealing the extraordinary odyssey of Homo sapiens. From our origins in the depths of prehistory to the intricate tapestry of modern civilization, this narrative is a testament to our species' remarkable journey.

It is a tale marked by adaptation, innovation, and an insatiable curiosity that sets us apart as a distinct branch on the tree of life. As we delve into this fascinating journey, we encounter the twists and turns of evolution that have sculpted us into the beings we are today.

Theories and Concepts

Lamarckism:

Jean-Baptiste Lamarck, a French biologist, put forth one of the earliest concepts of evolution during the late 18th and early 19th centuries. His theory, often termed "Lamarckism" or "the inheritance of acquired characteristics," played a role in the early stages of evolutionary thinking. However, it has lost credibility in modern biology primarily because it lacks substantial supporting evidence.

Lamarck's theory suggested that an organism could develop new characteristics or alter existing ones by using or not using specific body parts or organs. These changes, he argued, could then be inherited by the organism's offspring. Lamarck also proposed that organisms had a natural inclination to become complex and better adapted to their environments over time. He believed that this drive for complexity led to the development of more advanced species from simpler ones.

To illustrate his theory, Lamarck often used the example of giraffes. He proposed that giraffes had evolved long necks by stretching their necks to reach high leaves on trees. According to Lamarck, this stretching of their necks, acquired over their lifetimes, would be inherited by their offspring, resulting in giraffes with even longer necks in future generations.

Lamarck's ideas were based on limited scientific knowledge and lacked substantial empirical evidence and mechanisms. This theory has largely been relegated to the history of science, although it continues to be studied as an important historical precursor to the development of modern evolutionary theory.

Darwinism:

The British naturalist Charles Darwin was the pioneer in providing empirical evidence to explicate the possible mechanisms behind evolution through natural selection. His groundbreaking theory, commonly recognized as Darwin's Theory of Evolution, wrought a profound transformation in the field of biology as it offered a comprehensive explanation for the origins of the human species.

In the year 1831, when Darwin was a young 22-year-old student at the University of Cambridge, he was presented with a remarkable opportunity. As a budding naturalist, he embarked on a momentous expedition that spanned approximately five years, commencing in South America and encompassing multiple continents. During this expedition, Darwin collected a plethora of live specimens, meticulously crafted illustrations, and invaluable fossils from the regions he visited. These treasures he gathered played a pivotal role in laying the initial foundation for his profound comprehension of the intricate mechanisms governing the process of evolution.

During his explorations, Charles Darwin encountered the mylodon, a giant creature resembling a sloth. Intrigued by the interesting similarities between these seemingly diverse creatures, he began contemplating potential connections among them. However, it was on the Galápagos Islands in Ecuador that Darwin's insights truly flourished. There, he meticulously observed the tortoises inhabiting these distinct isles.

Remarkably, the tortoises showcased notable adaptations to their specific environments. On islands blessed with an abundance of low-lying vegetation, these creatures possessed shells contoured in a way that allowed them to effortlessly extend their heads downward, facilitating easy feeding. In stark contrast, tortoises dwelling on islands characterized by taller foliage sported dome-shaped shells, ideally suited for extending their necks upward to graze upon the towering vegetation.

These incisive observations catalyzed Darwin's observations regarding the complex interplay between environmental factors and the adaptive mechanisms of species. They sowed the seeds of his profound ideas on the evolution of life, ultimately concluding in his groundbreaking theory of natural selection.

Two decades following his successful voyage, Charles Darwin found himself in possession of writings, comprising thousands of pages of scientific investigation and observation. But he hesitated to unveil his ideas to the world because he wanted to have irrefutable evidence. Darwin was acutely aware of the seismic impact his theory would inevitably have on the scientific community and beyond.

Eventually, in the year of 1858, Darwin took the bold step of presenting his research to the public. It was in the subsequent year, 1859, that he published his seminal work, "On the Origin of Species." This momentous publication heralded a new era in the realm of scientific thought, propelling Charles Darwin to the apex of intellectual and scientific acclaim. His profound insights into the mechanisms of evolution forever carved his name into the annals of history as one of the most influential scientists to have ever graced the Earth.

Charles Darwin's theory of evolution suggests that species evolve through a process known as natural selection. Darwin proposed that within a given population, individuals with advantageous traits harmonizing with the demands of their environment enjoy a heightened likelihood of survival and reproduction. Consequently, passing those favorable traits on to their offspring. Through the passage of innumerable generations, this selective process leads to the gradual adaptation and modification of species, ultimately resulting in the diversification and emergence of new species. Central to his theory is the idea that all life shares a common ancestry, and the diversity of life on Earth can be explained by the branching and divergence of species over immense spans of time.

Neo-Darwinism:

Neo-Darwinism, also known as the modern synthesis, is a term used to describe the integration of Charles Darwin's theory of evolution by natural selection with Mendelian genetics.

Mendelian genetics, introduced by Gregor Mendel in the mid-19th century, encompasses the fundamental principles of heredity, where traits are passed from one generation to another. These principles include dominance and recessive-ness (certain traits masking others), segregation (the separation of alleles during gamete formation), and independent assortment (genes on different chromosomes assorting independently). This theory primarily applies to single-gene traits with two alleles, exemplified by Mendel's pea plant experiments. However, it's important to note that not all traits follow these straightforward patterns, as some are influenced by multiple genes and exhibit more complex inheritance patterns, expanding our understanding of genetics beyond Mendel's foundational work.

Neo-Darwinism upholds the central concept proposed by Charles Darwin, which suggests that species undergo an evolutionary process driven by natural selection. A fundamental aspect of this theory lies in the recognition of genetic diversity within populations. Genetic variation emerges as a result of mutations in an organism's DNA, leading to the formation of various gene variants known as alleles.

These genetic distinctions contribute to variations in traits among individuals within a species. Neo-Darwinism incorporates Mendelian genetics into this framework by acknowledging that genes, situated on chromosomes, act as carriers of genetic information governing an organism's traits. The different alleles at specific gene locations can undergo the forces of natural selection, ultimately enabling the evolution of traits across generations.

Population genetics is a crucial component of Neo-Darwinism, as it explores the alteration of gene frequencies within populations across time. It examines factors such as genetic drift, which entails random changes in gene frequencies, and gene flow, which involves the transfer of genes between different populations. These processes are studied alongside the concept of natural selection. Collectively, these mechanisms contribute to molding the genetic composition of populations and steering the course of evolutionary transformations.

Neo-Darwinism generally upholds the principle of gradualism, positing that evolutionary transformations unfold slowly and step by step across extended epochs. In contrast, alternative notions like punctuated equilibrium contend that certain changes may emerge swiftly during isolated episodes. Nonetheless, Neo-Darwinism underscores the accumulative character of evolutionary mechanisms.

A fundamental principle within Neo-Darwinism revolves around the concept of common ancestry. This notion has been supported by a wealth of molecular and fossil evidence. Furthermore, this theory highlights the significance of adaptation, emphasizing that living organisms undergo evolutionary changes to better suit their surroundings through the mechanism of natural selection. Characteristics that confer an advantage to an organism's survival and reproductive capabilities tend to become increasingly common within a population over successive generations.

In brief, Neo-Darwinism represents a fundamental framework in modern biology, offering a comprehensive understanding of how species evolve through genetic variation and natural selection. Although it remains a cornerstone of the study of evolution, it's important to acknowledge that the field is continually advancing, with ongoing investigations and findings further enhancing our grasp of the intricate mechanisms driving the diversity of life.

Punctuated Equilibrium:

Punctuated equilibrium is a theory in the field of evolutionary biology that proposes an alternative perspective to the conventional notion of gradual and continuous evolutionary change. This theory, first proposed by paleontologists Stephen Jay Gould and Niles Eldredge in the early 1970s, suggests that evolutionary development often occurs in relatively rapid bursts of change followed by long periods of stability.

According to the theory, species are believed to undergo relatively prolonged stable periods, during which they remain unchanged (referred to as "stasis"). These stable phases are occasionally punctuated by shorter intervals of rapid evolution or speciation. During these rapid evolutionary episodes, the emergence of new species can occur relatively swiftly, leading to significant shifts in characteristics and adaptation.

This idea presents a challenge to the traditional Neo-Darwinian notion of gradualism, which proposed that species undergo slow, continuous evolution over extended time spans. Punctuated equilibrium, in contrast, proposes that most species remain relatively stable for prolonged periods and only occasionally undergo rapid bursts of evolutionary change. These bursts are often triggered by environmental shifts or other influential factors.

One of the classic examples used to illustrate punctuated equilibrium is the fossil record of trilobites, a now-extinct assemblage of marine arthropods that thrived from the Cambrian period (approximately 541 to 485 million years ago) through the Permian period (approximately 298 to 252 million years ago). Trilobite fossils show episodes in which a species emerges in the geological record, remains substantially unchanged over the course of millions of years, and then experiences an abrupt disappearance, typically replaced by a novel species with distinct features. This phenomenon showcases the intermittent bursts of evolutionary change amid extended periods of species stability that punctuated equilibrium seeks to explain.

It's important to note that punctuated equilibrium is not intended to replace the idea of gradualism but rather offers an additional perspective on the pace and pattern of evolutionary transformation. Both gradualism and punctuated equilibrium are considered credible models for understanding how species evolve, and the relative importance of each may vary depending on the specific context and the species being studied.

Creationism:

Creationism, in relation to human evolution, is a worldview that posits a divine, supernatural origin for human beings, distinct from the evolutionary processes proposed by scientific theories. Rooted in religious and cultural narratives, creationism asserts that humans were intentionally created by a deity or deities according to specific beliefs and traditions. The narratives vary across different cultures and religions, each offering its unique account of how humans came into existence.

Creationist views also extend to notions of morality, purpose, and the nature of human existence. The concept of a divine creator, according to creationist beliefs, provides a framework for understanding the origin of moral values, the significance of human life, and the purpose for which humans were created. It offers a narrative that transcends the purely physical aspects of evolution, delving into metaphysical questions about the nature of consciousness, free will, and the human spirit.

One of the prominent forms of creationism, Young Earth Creationism (YEC), contends that the human species, along with all other forms of life, was created by a supernatural entity in a relatively brief period, typically within the span of a few thousand years. Young Earth

Creationists often interpret religious texts, such as the Bible, in a literal manner, asserting that the Earth's age is around 6,000 to 10,000 years.

Old Earth Creationism (OEC) takes a somewhat different stance, accepting the scientific age of the Earth while still proposing divine intervention or guidance in the process of human evolution. OEC attempts to reconcile certain aspects of evolutionary theory with a belief in a purposeful creation.

Intelligent Design (ID), another form of creationism, focuses on specific biological structures, arguing that certain features are too complex to have evolved gradually and must have been designed by an intelligent agent.

Intelligent Design:

Intelligent Design (ID), a notion that has ignited intense debate and reflection, aims to harmonize the intricacies of the natural world with the notion of deliberate design by an intelligent entity. While supporters of ID argue that certain attributes of living beings cannot be adequately explained through purely natural means, it is crucial to recognize that the scientific community generally regards ID as lying beyond the scope of mainstream science.

In the quest to comprehend Intelligent Design, one often encounters the work of Michael Behe, a biochemist whose contributions have fueled the ID discourse. Central to Behe's argument is the concept of "irreducible complexity." He posits that certain biological structures are so intricately interdependent that they could not have evolved gradually through small, successive changes. Instead, Behe suggests that these structures must have been designed in their entirety by an intelligent agent.

Behe's seminal work, "Darwin's Black Box," published in 1996, provides a comprehensive exploration of the idea of irreducible complexity. He delves into the intricate machinery of the cell, pointing to biochemical systems that, in his view, defy explanation through gradual evolutionary processes. The bacterial flagellum, a whip-like appendage used for propulsion in certain bacteria, is a frequently cited example. Behe contends that the flagellum's structure is irreducibly complex, meaning that the removal of any one part would render it nonfunctional. This, he argues, challenges the conventional understanding of evolutionary development.

Phillip E. Johnson, a law professor, is another influential figure associated with the promotion of Intelligent Design. His work, particularly "Darwin on Trial," has sought to challenge the foundations of evolutionary theory. Johnson critiques the adequacy of natural selection in explaining the complexity of life, presenting arguments that echo broader concerns about the limitations of purely naturalistic explanations.

In the broader context of philosophy and theology, Intelligent Design resonates with those who find in it a bridge between scientific inquiry and a belief in a purposeful universe. It addresses questions about the origins of life and the intricacies of biological systems, providing an alternative narrative to the naturalistic explanations put forth by evolutionary biology.

Genetics:

Genetics is the scientific study of heredity, exploring the mechanisms by which traits are passed down from one generation to the next in living organisms. At the core of genetics is the concept of genes, which are segments of DNA (deoxyribonucleic acid) containing instructions for building and maintaining the structures and functions of cells. These genes serve as the blueprint for the development, growth, and functioning of an organism. The entire collection of an organism's genes is referred to as its genome.

The study of genetics involves understanding how genes are transmitted during reproduction, how they interact with each other, and how they can undergo changes over time. Gregor Mendel's groundbreaking work in the 19th century laid the foundation for modern genetics by establishing the principles of inheritance. The field has since evolved with the discovery of the structure of DNA by

James Watson and Francis Crick, further advancing our understanding of the molecular basis of genetics. Molecular genetics delves into the intricate processes of gene expression, replication, and regulation at the molecular level, providing insights into the dynamic interplay of genetic information within cells.

Genetics plays a crucial role in various aspects of biology, from unravelling the mysteries of inherited diseases to exploring the evolutionary relationships among species. Recent advances in genetic technologies, such as DNA sequencing and gene editing, have revolutionized the field, offering unprecedented tools for studying genes and their functions. The study of genetics not only enhances our comprehension of individual traits but also plays a crucial role in broader investigations into the origins, diversity, and evolution of life on Earth.

Genetic investigations into human evolution involve the analysis of ancient DNA from archaeological remains, providing a molecular window into the past. By examining genetic markers and comparing DNA sequences across populations, researchers can reconstruct the migratory patterns, interbreeding events, and adaptive changes that define human evolution. Population genetics plays a crucial role in understanding the dynamics of genetic variation within and among human populations, offering insights into factors like natural selection, genetic drift, and gene flow that have influenced our genetic landscape.

RNA, a close cousin to DNA, also plays a crucial role in human evolution. RNA stands for Ribonucleic Acid. It is a type of nucleic acid. While DNA carries the long-term genetic instructions, RNA acts as an intermediary messenger, facilitating the translation of genetic information into proteins—the building blocks of cells. The study of RNA, particularly in the context of gene expression and regulation, helps elucidate the dynamic processes that contribute to the complexity of human development. The emerging field of epigenetics, which involves modifications to DNA and associated proteins, further underscores the role of RNA in mediating environmental influences on gene expression and, consequently, human evolution. Genetics, encompassing both DNA and RNA, serves as a powerful tool for unravelling the intricate story of human evolution, offering glimpses into our shared ancestry and the forces that have shaped our species over millennia.

Advancements in genomics have further amplified our understanding of human evolution by allowing for the comprehensive study of entire genomes. Comparative genomics reveals shared genetic elements across species, elucidating our genetic connections with other primates and offering clues about the genomic innovations that set humans apart. Additionally, studies of specific genes associated with traits like language development, cognitive abilities, and resistance to diseases provide glimpses into the selective pressures that have shaped the unique features of the human genome. Genetics provides a molecular compass guiding the exploration of the intricate mosaic of human evolution, offering profound insights into our shared ancestry and the genetic tapestry that defines us as a species.

Micro Evolution and Macro Evolution:

Microevolution and macroevolution are two interconnected concepts in evolutionary biology that describe different scales and scopes of the evolutionary process.

Microevolution refers to the small-scale changes in the genetic makeup of populations within a species over relatively short periods of time. It involves processes such as natural selection, genetic drift, gene flow, and mutation. These mechanisms lead to changes in the frequency of alleles (different versions of a gene) within a population. Microevolutionary changes can be observed within a species and are often associated with adaptations to local environments, changes in physical traits, and variations in behaviors. Classic examples include the evolution of antibiotic resistance in bacteria or the adaptation of Darwin's finches to different food sources on the Galápagos Islands.

Macroevolution, on the other hand, takes a broader view and addresses large-scale patterns and processes in evolution. It focuses on the evolution of entire taxonomic groups, the origin of new species, and the development of major evolutionary trends over geological time scales. Macroevolutionary events are responsible for the diversification of life, including the emergence of new body plans, the origin of major groups of organisms, and mass extinctions. Examples of macroevolutionary phenomena include the radiation of mammals after the extinction of dinosaurs, the development of complex organs like eyes, and the origin of different animal phyla during the Cambrian explosion.

While microevolution and macroevolution are conceptually distinct, they are deeply interconnected. Microevolutionary processes occurring within populations contribute to the larger patterns and changes observed at the macroevolutionary level. Together, these concepts provide a comprehensive framework for understanding the complexities of evolutionary processes that have shaped life on Earth over millions of years.

Origin of Life:

The concept of the origin of life is distinct from the theory of evolution, although they are interconnected in the broader context of biology. The origin of life addresses the foundational question of how life was initiated on Earth, while evolution explores the subsequent transformations and adaptations that occurred in living organisms once life was established. The origin of life can be viewed as a narrative written in the language of chemistry, where the evolution of chemicals set the stage for the mesmerizing journey from inanimate matter to the dazzling complexity of life forms we observe today.

While the question of how and when life was initiated on our planet is a complex scientific inquiry, two principal theories take the forefront in this discourse.

Abiogenesis or Prebiotic Chemistry: Abiogenesis is the scientific hypothesis proposing that life originated from non-living matter through an intricate sequence of chemical reactions that spanned billions of years.

According to this theory, the Earth's early environment, approximately 3.5 to 4 billion years ago, provided the ideal conditions for the creation of basic organic molecules. These molecules included essential building blocks such as amino acids and nucleotides, which formed through various processes such as chemical reactions occurring in hydrothermal vents, within the atmosphere, or on the surfaces of minerals. Over extended periods, these fundamental organic compounds had the potential to combine and undergo transformations, leading to the emergence of more intricate molecular structures. Eventually, this process is believed to have given rise to the very first simple life formation of simple single-celled organisms.

In 1953, Stanley Miller and Harold Urey conducted a famous experiment that simulated the conditions believed to exist on early Earth. They exposed a mixture of water, methane, ammonia, and hydrogen to electrical sparks to mimic lightning. This experiment produced amino acids, which are the building blocks of proteins, supporting the idea that complex organic molecules could have formed under such conditions.

Some researchers propose that life could have originated around hydrothermal vents on the ocean floor. These vents release high-temperature, mineral-rich fluids into the surrounding seawater. The chemical gradients at these vents, along with the energy they provide, might have fostered the formation of organic molecules and the emergence of early life.

Another hypothesis suggests that certain minerals, particularly clay minerals, may have played a role in the formation and concentration of organic molecules. These minerals could have acted as a template for the assembly of simple organic compounds.

Despite being a widely accepted theory, the specific processes that led to the transition from non-existence to life continue to be a subject of active scientific investigation. Researchers have proposed alternative theories and are persistently exploring the intricate details of how the initial organic molecules emerged and evolved.

Panspermia: Panspermia is a hypothesis that presents an intriguing perspective on the origin of life on Earth. Instead of proposing that life originated here, it suggests that life did not have its genesis on our planet but was, in fact, transported to Earth from other regions of the universe. This transport could have occurred through various means, including the delivery by comets, asteroids, or cosmic dust particles. According to this theory, the fundamental components of life or even basic microorganisms may have pre-existed beyond Earth and embarked on a journey to reach our planet.

It's important to note that while panspermia offers an alternative viewpoint on the distribution of life, it does not provide an explanation for the initial emergence of life itself. Rather, it posits that life may have existed in other extraterrestrial environments before making its way to Earth. This concept raises intriguing questions about the potential existence of life beyond our world and the mechanisms by which it could have travelled through space.

Support for the panspermia hypothesis comes from various sources. Scientists have detected organic molecules in space, demonstrating that the building blocks of life are not exclusive to Earth. Additionally, the resilience of certain microorganisms in harsh space conditions, including extreme temperatures and radiation, suggests that life could potentially survive the journey between celestial bodies. These observations contribute to the ongoing exploration of panspermia as a fascinating avenue in the quest to understand the origins and distribution of life in the universe.

These theories are not mutually exclusive, and some scientists propose that life on Earth could have multiple origins or may have been influenced by extraterrestrial sources. Furthermore, the study of life's origin is challenging due to the lack of direct evidence from billions of years ago.

Last Universal Common Ancestor (LUCA):

The Last Universal Common Ancestor, often abbreviated as LUCA, is a concept within the realms of evolutionary biology and microbiology. It represents the theoretical single-celled organism that is considered to be the origin point from which the three fundamental domains of life emerged: Bacteria, Archaea, and Eukarya. In simpler terms, it is believed that every form of life on our planet can trace its lineage back to LUCA. It is thought to be the most recent common ancestor of all living organisms, and it is estimated to have existed billions of years ago, during the infancy of Earth.

LUCA is a pivotal concept in the study of the Tree of Life and the origin of species. Even though LUCA itself has never been observed or identified. Scientists employ the tools of comparative genomics and phylogenetic analysis to glean insights into its probable characteristics and the environmental context in which it thrived. These studies suggest that LUCA was a simple micro-organism, possibly resembling a modern-day prokaryote (such as a bacterium or archaeon), and it would have lived in a very different world from the one we know today, likely in a high-temperature, anaerobic (oxygen-free) environment.

Some of the scientific evidence and methods used to infer LUCA include,

Comparative Genomics: One of the primary lines of evidence for LUCA comes from comparative genomics. By examining the genetic material (DNA and RNA) of various living organisms, scientists can identify common genes and genetic sequences that are shared among diverse life forms. The presence of these shared genes suggests a common ancestral organism, which is inferred to be LUCA.

Phylogenetic Analysis: It is a study of the evolutionary relationships between different species. By constructing phylogenetic trees based on genetic and molecular data, scientists can trace the evolutionary history of organisms. LUCA is considered the root of the tree of life, and its existence is inferred from the branching patterns of this tree.

Universal Biomolecules: Certain biomolecules and cellular structures are found in all known life forms on Earth. For example, the genetic code, the basic structure of DNA and RNA, and the fundamental machinery of protein synthesis are remarkably consistent across all living organisms. These shared features imply a common ancestry and provide evidence for LUCA.

Ribosomal RNA (rRNA): Ribosomal RNA plays a crucial role in protein synthesis in all living cells. The similarity of rRNA sequences across diverse organisms suggests a common ancestor from which this molecular machinery evolved.

Universal Metabolic Pathways: Many metabolic pathways are shared among different organisms, indicating a common origin. For example, the core metabolic processes involved in energy production and the utilization of organic compounds are conserved across various life forms.

Thermophilic and Anaerobic Characteristics: Some characteristics of LUCA are inferred from the types of environments and conditions in which early life is thought to have thrived. Based on geological and geochemical evidence, scientists believe LUCA may have lived in high-temperature, anaerobic environments, and this is deduced from the biochemistry of the shared ancestral traits of organisms.

It's important to note that while these lines of evidence strongly support the existence of LUCA, the exact nature and characteristics of this ancestral organism remain a subject of ongoing research and debate. As new scientific discoveries and advancements are made, our understanding of LUCA and the origins of life on Earth may continue to evolve.

Taxonomy of Human Evolution

The Taxonomy of Human Evolution intricately delves into the systematic classification of diverse hominid species, meticulously tracing their evolutionary relationships. This comprehensive classification incorporates an array of evidence, ranging from fossils to archaeological findings and genetic studies, aiming to unravel the intricate tapestry of human evolution. Through this methodical taxonomy, scientists seek to organize and categorize the various species that have shaped the evolutionary journey leading to modern humans. This scholarly pursuit not only illuminates the branching points but also sheds light on the intricate relationships among different hominids spanning across geological epochs.

The primary levels of taxonomy, from the most general to the most specific, are as follows:

Domain: This is the broadest category and represents the highest level of classification. There are three domains: Bacteria, Archaea, and Eukarya. Bacteria and Archaea are primarily composed of unicellular microorganisms. Humans belong to the domain *Eukarya*, which includes all organisms with complex, membrane-bound cells (eukaryotic cells).

Kingdom: Below the domain level, organisms are grouped into kingdoms based on certain shared characteristics. For example, in the Eukarya domain, some common kingdoms include Animalia (animals), Plantae (plants), Fungi, and Protista (single-celled eukaryotes). Humans are classified in the kingdom *Animalia*, which comprises multicellular, heterotrophic organisms.

Phylum: Within each kingdom, organisms are further divided into phyla based on additional distinctive traits. Humans belong to the phylum *Chordata*, a group of animals characterized by the presence of a notochord at some stage in their development. In humans, the notochord develops into the vertebral column (spine).

Subphylum: Within the phylum Chordata, organisms are further divided and categorized based on additional shared characteristics and evolutionary relationships. Humans fall into the subphylum *Vertebrata*. This subphylum includes chordates with a well-developed vertebral column, which surrounds and protects the dorsal nerve cord.

Class: Phyla are subdivided into classes, which group organisms that share similar features. In the subphylum Vertebrata, humans are classified in the class *Mammalia*. Mammals are characterized by several distinctive features, including the presence of mammary glands that produce milk, hair or fur, and a warm-blooded metabolism.

Order: Classes are further divided into orders, which represent more specific groupings based on shared characteristics. For example, within the class Mammalia, the order Carnivora includes carnivorous mammals like cats and dogs. Within the class Mammalia, humans belong to the order *Primates*. Primates are a group of mammals that includes humans, apes, monkeys, and prosimians. Primates are known for their well-developed brains, grasping hands and feet, and typically forward-facing eyes.

Family: Orders are subdivided into families, which encompass related genera and species. In the order Primates, humans are classified within the family Hominidae, commonly known as hominids. Hominids include a broader range of species, encompassing all great apes such as orangutans, gorillas, chimpanzees, bonobos, and humans. Hominins, a more specific term, fall under one of the subcategories of the family, referring to the tribe that includes humans and our closest extinct relatives.

Genus: Families consist of genera (plural of genus), which are groups of closely related species. The genus name is often employed as the first part of an organism's scientific name. Within the family Hominidae, humans are classified in the genus *Homo*. This genus encompasses various extinct species of early humans, such as Neanderthals and Denisovans, alongside the sole surviving species, Homo sapiens.

Species: The most specific level of taxonomy is the species, which represents a group of organisms that can interbreed and produce fertile offspring. The species name for modern humans is ***Homo sapiens***. "Homo" refers to the genus, and "sapiens" is the species name. This classification distinguishes us as the only surviving species within the genus Homo.

This hierarchical system of taxonomy serves as a crucial framework, aiding scientists in the categorization and comprehension of relationships between various species, including humans. Grounded in shared characteristics and evolutionary history, this system offers a structured approach to classifying life forms. It is imperative to recognize that taxonomy, despite its seemingly static nature, is subject to evolution itself. The field can change over time, driven by the emergence of new discoveries, advancements in scientific methodologies, and a continually evolving comprehension of relationships among organisms.

Stages of Human Evolution

The trajectory of human evolution is delineated through various species and subspecies, offering insights into the evolutionary journey leading to Homo sapiens, the modern humans. These classifications, derived from the fossil record and genetic evidence, remain dynamic, subject to modification with each new discovery. The following outlines only some key stages in the continuum of human evolution,

Sahelanthropus tchadensis (7 to 6 million years ago): It holds a special place in the annals of human evolution as one of our earliest known hominins. Fossil remnants of Sahelanthropus were unearthed in Chad, Africa, shedding light on this ancient relative of ours. What makes Sahelanthropus particularly intriguing is its blend of characteristics, displaying a combination of features reminiscent of both apes and early humans.

Ardipithecus (around 4.4 million years ago): Ardipithecus ramidus, commonly known as "Ardi," is one of the earliest known hominins. Ardi's existence traces back to the era when early hominins were in the nascent stages of emergence. Ardi's choice of habitat leaned toward forested areas, suggesting a preference for wooded environments.

What sets Ardi apart as a fascinating subject of study is the combination of traits they possessed. Some of these traits resembled those of the distant ape ancestors, indicating that they retained certain primitive features. However, Ardi also exhibited traits that hinted at the early stages of human development, providing valuable insights into the transitional phase of our evolutionary history.

Australopithecus (around 4 to 2 million years ago): The Australopithecus genus comprises widely recognized fossil specimens, with "Lucy" standing as a prominent example identified under the classification Australopithecus afarensis. These early hominins demonstrated the remarkable trait of bipedalism, signifying their ability to walk upright on two legs—a crucial evolutionary development. Their physical characteristics showcased an intriguing amalgamation of both ape-like and human-like attributes.

The discovery of Australopithecus fossils traces back to 1924, with the initial findings in South Africa. These creatures were relatively diminutive, standing at approximately 4 feet in height and weighing between 60 to 80 pounds. As they adapted to life on the ground, Australopithecus individuals began utilizing stone tools, indicating a rudimentary form of tool used for various purposes, including defense and sustenance. Moreover, their posture evolved to a more erect stance while walking, setting the stage for further advancements in human locomotion and adaptability.

Paranthropus (around 2.7 to 1 million years ago): Paranthropus species, among them Paranthropus robustus and Paranthropus boisei, possessed robust cranial structures and teeth that were well-suited for consuming plant-based diets. These remarkable adaptations enabled them to thrive on vegetation. Notably, they shared their environment with early Homo species, implying that different hominin branches coexisted during this period, each adapted to its own specific ecological niche.

Homo habilis (around 2.4 to 1.4 million years ago): Homo habilis, often referred to as the "Handy Man", is regarded as one of the initial members of the Homo genus. This species distinguished itself through its remarkable ability as toolmakers, crafting instruments that served various purposes. Additionally, Homo habilis displayed a notable advancement in cranial capacity compared to their predecessors among hominins. This increase in brain size was a significant step in the evolution of our lineage, reflecting an enhanced cognitive capacity and adaptation.

Homo erectus (around 1.9 million to 140,000 years ago): Homo erectus displayed a body structure that bore a closer resemblance to modern humans. They demonstrated a notable advancement in tool-making, indicating a greater level of technological sophistication. Moreover, they hold the distinction of being the earliest hominin species to embark on a significant migration, venturing beyond the confines of Africa into the vast expanses of Eurasia.

What sets Homo erectus apart is their ability to walk exclusively on two legs, a form of locomotion known as bipedalism. This adaptation not only marked a crucial step in our evolutionary journey but also distinguished them as one of the first hominin species to consistently adopt this mode of movement.

Homo heidelbergensis (around 700,000 to 200,000 years ago): Homo heidelbergensis is believed to be the shared ancestor of both Neanderthals and modern humans. These individuals possessed a larger cranial capacity, indicative of a relatively larger brain, and displayed a notable advancement in toolmaking compared to their predecessors. This suggests a significant step forward in the cognitive and technological abilities of our human lineage.

Neanderthals (around 400,000 to 40,000 years ago): Neanderthals constituted a separate and closely linked species to modern humans. They possessed a robust physical build, demonstrating considerable strength and adaptability. What sets them apart is their ability to craft advanced tools, indicative of their cognitive capabilities. Beyond their technical prowess, Neanderthals displayed a rich and intricate culture, which included social structures, rituals, and possibly symbolic expressions. They thrived primarily across the vast expanse of Eurasia, adapting to various ecological niches and climatic challenges.

Neanderthals are believed to have coexisted with early modern humans (Homo sapiens) for a significant period before eventually disappearing from the fossil record. The exact reasons for their extinction are still a subject of scientific investigation and debate. Interestingly, the DNA inherited from Neanderthals can still be found in the genetic makeup of many people today. Modern humans of non-African descent carry a small percentage (typically around 1-2%) of Neanderthal DNA in their genomes.

Homo sapiens (modern humans, approximately 300,000 years ago to present): Homo sapiens, as widely believed, originated in Africa and subsequently dispersed to various regions across the globe. Modern humans exhibit distinctive anatomical features, including a prominent forehead, a well-rounded skull, a pronounced chin, and a relatively large brain in proportion to their body size.

Fossil Discoveries

A fossil is like a time capsule from the past. It is the preserved remains, traces, or impressions of a pre-existing organism or evidence of ancient life that has been preserved in rocks, minerals, or other geological materials over long periods of time. Fossils can include the remains of plants, animals, and even microorganisms. They provide valuable insights into the history of life on Earth, including information about extinct species, their anatomy, behavior, and the environments in which they lived. Fossils can take various forms, such as bones, teeth, shells, imprints, tracks, and petrified wood, and they are typically found in sedimentary rocks where the conditions for preservation were favorable.

For the successful formation of a fossil from the remains of an organism, certain specific conditions must be met. Firstly, it requires

The Quranic narrative on Human Evolution

that the deceased organism be spared from immediate consumption by scavengers and be situated in an environment where oxygen levels are low, thus impeding rapid decomposition. If scavengers have already had their share, a substantial portion of the organism's remains must endure, ensuring the necessary conditions for fossilization. Subsequently, the remains are interred within layers of sediment, such as sand or mud. Over time, additional layers of sediment accumulate, gradually solidifying into what is recognized as sedimentary rock. Concurrently, minerals present in the surrounding water gradually infiltrate the porous structure of the bones. Through a process known as mineral replacement, these minerals gradually supplant the original bone material, effectively transforming the entire fossilized relic into stone.

To provide a rough estimate, it is noteworthy that, on average, several dozen to a few hundred early human fossils are uncovered and documented annually. The discovery of early human remains provides researchers with a greater wealth of evidence to construct the narrative of our origins. The term "early humans" encompasses a diverse array of hominid species, including Australopithecus, Homo erectus, Homo habilis, and Neanderthals, among others. Presented below are a few illustrative examples of such fossils.

Graecopithecus discovered in Greece & Bulgaria, around 7 Million Years Old

In the year 1944, an intriguing discovery unfolded in the historic city of Athens, Greece—a lower jawbone that would beckon the curiosity of scientists and stir the annals of human evolution. Employing the advances of modern dating technology, the age of this remarkable artefact was estimated to be a staggering 7.24 million years old. What made this discovery even more interesting was its uncanny resemblance to a single upper premolar tooth unearthed in Bulgaria.

These remains were ascribed to Graecopithecus, which is believed to be an early human ancestor. Scientists reached this conclusion by studying the jaw and tooth, noticing a key difference that the roots of the molar were partly fused together, a distinctive characteristic exclusive to human dental morphology. In stark contrast, the roots of the apes, typically have three or four roots that spread apart.

The fusion of roots and other characteristics rendered compelling evidence that the jawbone belonged to a hominid, kin to our ancestral line. Yet, as with many profound discoveries, this revelation did not unfold without controversy. Some dissenting voices among paleontologists suggest an alternative narrative that Graecopithecus may not be a forebear to humankind at all. They contend that it might be more aptly classified as a relic of an ancient ape lineage known as Ouranopithecus. They argue that relying solely on a single tooth and jawbone as evidence is insufficient to conclusively establish a human origin, and thus, the mystery endures.

Sahelanthropus discovered in Chad (Africa), Around 7 Million Years Old

In the year 2001, an important discovery unfolded in Chad, located in Central Africa. This remarkable find was made possible through the efforts of French paleontologist Michael Brunet and his team of experts, who embarked on extensive excavations that yielded an array of skulls and jawbones.

Detailed analysis by these experts revealed distinctive characteristics of these early humans. They possessed comparatively small brains, elongated skulls, and pronounced brow ridges. The manner in which their spinal cords connected to the skull proved particularly enlightening, prompting scientists to deduce that Sahelanthropus possessed the capability of bipedal locomotion, or walking upright. The dating of these fossils places them at an age of approximately 7 million years, a bit later than the Graecopithecus discovery by about a hundred thousand years. However, the geographical location of these fossils in Central Africa aligns with the widely accepted theory that points to Africa as the cradle of human origins.

Ardipithecus kadabba discovered in Ethiopia, Around 5.2 to 5.8 Million Years Old

Between 1997 and 2004, significant human fossils were found in Ethiopia's Awash River valley, causing quite a stir among archaeologists. An Ethiopian paleontologist uncovered a portion of a lower jaw along with fragments of teeth, toes, arms, and collar bones. These remains were believed to belong to a hominid species. Remarkably, the discovered fossils were estimated to be from a period approximately 5.6 to 5.8 million years ago. Notably, among these remains, a specific toe bone was found to be even more ancient.

Subsequently, eleven additional specimens were discovered, ultimately leading to the conclusion that a completely new species of human ancestor was discovered. This species exhibited a bipedal walk, possessed sizable molars, and featured a relatively small brain, resembling apes more than hominids. These primitive, ape-like characteristics prompted certain scientists to raise the question of whether Ardipithecus kadabba should be categorized as a human ancestor or an ape.

Lucy discovered in Ethiopia Around 3.2 Million Years Old

In the year 1974, a groundbreaking discovery unfolded in Hadar, Ethiopia—the unearthing of Lucy, the oldest and most complete pre-human fossil ever found. Remarkably, Lucy's remains constituted 40% of an entire human skeleton. Lucy's age was estimated to be 3.2 million years old, and she was identified as a hominid with clear evidence of bipedalism, meaning she walked on two legs.

Further investigations revealed that Lucy was an adult female at the time of her demise. Scientists went so far as to reconstruct an image of what she might have looked like in life. Even in contemporary times, Lucy remains one of the most renowned human fossils, often regarded as a crucial piece of evidence linking us to our hominid ancestors and a common, ape-like forebear.

Neanderthal 1 discovered in Germany, Around 400,000 Years Old

It was within the recesses of the Feldhofer Cave, nestled in the Neander Valley of Germany, that the first fossil of Neanderthal was unearthed in 1856. These remnants revealed a distinctive oval-shaped skull with a receding, sloped forehead, pronounced brow ridges, and robust, sturdy bones.

Neanderthals stand as our oldest known extinct human relative. Their legacy is interwoven with the tapestry of European history, for they are the forebears of modern humans on the European continent. Neanderthals inhabited a considerable period, ranging from 400,000 to 40,000 years ago.

In their quest for sustenance, Neanderthals pursued a diet that encompassed both plant and animal fare. They wielded sharp spears for hunting, securing meat that was crucial for their survival during harsh winters. Additionally, evidence derived from the fossil record hints at their pioneering ingenuity—the crafting of clothing from animal skins, marking them as our ancestors with the distinction of being the first to do so.

Yet, it is their complex behaviors that bring Neanderthals closest to modern humans. Notably, they exhibited a form of ritualistic respect for their deceased, engaging in the intentional burial of their dead, a practice that underscores the profound connections between their world and ours.

Jebel Irhoud Fossils discovered in Morocco, Around 300,000 Years Old

In the year 2017, a momentous discovery unfolded amidst the mountains of Morocco—a cache of fossils that defied expectations, dating back to 315,000 years. This discovery sent ripples through the scientific realms of paleontology and archaeology, shaking up long-held beliefs.

The sheer antiquity of these fossils was a profound surprise, surpassing the timeline envisioned by scientists. Furthermore, their

The Quranic narrative on Human Evolution

location in Morocco stood far afield from Central Africa, where the origins of modern humans were traditionally thought to lie. Up until then, the prevailing wisdom had led scientists to anticipate that older remains of Homo sapiens would exclusively surface in East Africa, which had been regarded as the cradle of humankind.

Omo Remains discovered in Ethiopia, Around 200,000 Years Old

In addition to the remarkable finds in Morocco, some of the earliest fossils revealing signs of the transition to modern humans were uncovered in Ethiopia, dating back between 195,000 and 233,000 years ago.

One particularly significant discovery emerged in the form of a skull, dating back 195,000 years, found within Ethiopia's Omo-Kibish Formation. Intriguingly, there's a possibility that these Omo fossils could be even older—some scientists suggest they might extend as far back as 233,000 years, based on more recent and precise dating methods applied to volcanic ash in the region.

Mungo Man discovered in Australia, Around 42,000 Years Old

In the Lake Mungo Region of New South Wales, Australia, two sets of human-remain, known as Mungo Man and Mungo Lady, were discovered in the years 1969 and 1974. These remains were found at approximately 1600 feet from each other and suggested careful burial, possibly with ceremonial significance. Notably well-preserved, they provided a compelling glimpse into the bygone era.

Through meticulous examination, scientists were able to glean insights about Mungo Man. They determined that he was likely a male who had reached the age of 50, bearing signs of osteoarthritis. It's also believed that he might have been involved in activities such as basket weaving or crafting fishing nets.

The area surrounding these remains yielded further evidence of human occupation, including hearths, grindstones, stone tools, and discarded remnants from meals that included fish, mussels, fowl, and eggs. These discoveries provide a vivid portrayal of the lives of ancestral

Aborigines and significantly challenge previous assumptions about the timing of their occupation of the land, indicating a much earlier presence than commonly believed.

Tianyuan Man discovered in China around 39,000 to 42,000 Years Old

The discovery of the Tianyuan skeleton in Tianyuan Cave, located in Zhoukoudian, Beijing, in 2003 marked a significant milestone in the study of early humans. These fossils were among the earliest evidence of humans living in Eurasia, with radiocarbon dating estimating their age at approximately 39,000 to 42,000 years old.

Tianyuan Man was unequivocally a Homo sapiens, belonging to the lineage of our Asian forebears. Within the cave, researchers unearthed thirty-four bone fragments that provided insights into this ancient individual. However, notably absent were any other artefacts or remnants that could shed light on the man's way of life, dietary preferences, or cultural practices.

Cro-Magnon Man discovered in France, Around 30,000 Years Old

Among the old fossils of early modern humans lies a discovery from the year 1868, nestled in the southwestern region of France, within the renowned village of Les Eyzies. These ancient people are famously known as Cro-Magnon Men and were of height, ranging from 5 feet 4 inches to around 6 feet tall.

Following meticulous excavation efforts, the remnants of five individuals were uncovered, alongside a treasure trove of artifacts: stone tools, ivory pendants, seashells, and intricately carved reindeer antlers. The age of this site was determined to be approximately 30,000 years old, firmly establishing these as some of the most ancient Homo sapiens fossils ever unearthed in the European continent.

The discovery of the Cro-Magnon fossils was marked by a significant observation—their location and arrangement strongly suggested a purposeful burial. Among these fossils, there was a nearly complete skull belonging to a middle-aged male. This skull exhibited striking resemblances to the anatomy of modern humans.

Contribution of Muslim Scientists
Imam Al-Jahiz:

His full name was Abu Uthman Amr ibn Bahr al-Kinani al-Fuqaimi al-Basri a.k.a. Al-Jahiz (776–868 CE). He was a prominent Arab scholar, theologian, grammarian, and writer during the Islamic Golden Age. He is most celebrated for his magnum opus, "Kitab al-Hayawan" (The Book of Animals), a comprehensive zoological encyclopedia that comprises meticulous observations and profound discussions about the natural world. Within the pages of this remarkable tome, he articulated several ideas that can be identified as early influences on the developing field of evolutionary thought.

Observation of Variation and Adaptation: Al-Jahiz conducted careful observations and documented the diverse characteristics and adaptations of animal species to different environments. He incisively noted that animals exhibited varying physical traits and behaviors, often influenced by their specific habitats and survival requirements.

Discussion of Competitive Dynamics: He delved into the concept of competition among animals for essential resources like food and territory. He shrewdly recognized that certain animals possessed advantageous traits that better suited them for survival and reproduction. This notion foreshadows the concept of natural selection that would later become fundamental in evolutionary theory.

Consideration of Environmental Impact: Al-Jahiz proposed that environmental factors, encompassing variables such as climate and the availability of sustenance, could exert a transformative influence on the traits and characteristics of animals over time. He contended that these environmental factors could potentially give rise to alterations within species.

Ibn Miskawayh:

Abu Ali Aḥmad ibn Muhammad ibn Yaqūb ibn Miskawayh was a prominent Persian philosopher, historian, and scholar who lived during the 10th century (932-1030 CE). His significant contributions left an enduring mark on the evolution of Islamic philosophy and ethical thought. Of his philosophical opus, "Tahdhib al-Akhlaq" (The Refinement of Character) stands as a preeminent work, a testament to his profound exploration of ethics and moral philosophy within the Islamic intellectual tradition.

Ibn Miskawayh's ideas on evolution, while intriguing, are situated within the framework of philosophical and metaphysical reasoning, distinct from the empirical foundations of contemporary evolutionary theory. They bear the indelible imprint of the Aristotelian and Neoplatonic traditions that held sway in the Islamic philosophical milieu of his era. Within this context, he advanced a conceptual framework that can be distilled into several pivotal stages,

Mineral Evolution: Ibn Miskawayh posited a captivating notion that minerals, over time, could undergo a process of evolution. In his view, the simplest and most rudimentary minerals had the potential to gradually transform into more intricate and sophisticated forms. This concept parallels the earlier belief in the transmutation of elements, a prevailing notion within the realm of early natural philosophy.

Plant Evolution: Building upon the foundation of mineral evolution, Ibn Miskawayh extended his evolutionary schema to include plants. He articulated the idea that plants emerged as the subsequent stage in this evolutionary process. According to his vision, plants evolved from minerals, representing a heightened level of organization within the natural world.

Animal Evolution: He further broadened his evolutionary framework by proposing the evolutionary development of animals from plants. He contended that animals represented a higher level of complexity in comparison to plants, thus marking another progression in this philosophical narrative.

Human Evolution: In the culmination of his evolutionary perspective, Ibn Miskawayh postulated that humans occupied the pinnacle of this unfolding process. He regarded humans as possessing the most advanced and intricate organization among all living beings.

Ibn Khaldun:

Abu Zayd Abd al-Rahman ibn Muhammad ibn Khaldun stands as a luminous beacon in the annals of Islamic scholarship during the 14th century. His multifaceted contributions to history, philosophy, sociology, and the broader social sciences transcend boundaries, rendering him a revered figure within and beyond the Islamic world.

Central to his intellectual legacy is his magnum opus, "Al-Muqaddimah" (The Introduction), often referred to as "Ibn Khaldun's Prolegomena". This monumental work, acclaimed as the foundational cornerstone of historiography, sociology, and the philosophy of history, has left an indelible imprint on the realms of thought and inquiry.

Within the pages of this esteemed tome, Ibn Khaldun's profound ruminations extend to the concept of evolution, he explains the stages of evolution as,

Evolution of Minerals: Evolution started from minerals and progressed, in an ingenious, gradual manner to plants and then animals. Elaborating on the connection he stated that the last stage of each group is fully prepared to become the first stage of the next group.

Minerals to Plants Evolution: He states the last stage of evolution in minerals, connected with the first stage of plants, such as herbs and seedless plants.

Plants to Animals Evolution: He states the last stage of evolution in plants, such as palms and vines, connects to the first stage of animals, such as snails and shellfish which have only the sense of touch.

Animals to Humans Evolution: The animal world widened with numerous species and then in a gradual process of evolution, it finally reached to man, who can think and reflect. Ibn Khaldun intriguingly extends this idea by positing that the pinnacle of human development is attained through a progression from the realm of monkeys. In this perspective, he suggests a transition from the last stage of monkeys to the initial stages of humanity.

Four centuries before Darwin's epochal work, Ibn Khaldun established a thought-provoking link between the ancestry of humans and that of monkeys. Remarkably, there is no historical evidence to suggest that this concept ignited scandal during its time, yet it has endured as a contentious focal point of debate, serving as a divisive juncture between proponents and opponents of the Darwinian theory.

Although the observations and ideas put forth by these eminent Muslim scientists held substantial significance within the contexts of their respective epochs, it is imperative to regard their contributions as foundational rather than exhaustive when compared to the comprehensive framework of modern evolutionary theory.

It is crucial to acknowledge that these scholars did not formulate a fully-fledged theory of biological evolution. Nevertheless, their contributions to the understanding of the natural world stand as commendable achievements, particularly when considering the constraints imposed by limited resources for knowledge acquisition and empirical study during their times.

Chapter Three: Science vs Religion

The exploration of human evolution as a scientific subject is a relatively recent development in the broader context of modern science. Its emergence into mainstream discourse occurred in the mid-1800s, capturing the keen interest of both scientists and theologians alike. However, this emergence also brought with it a significant divergence of perspectives, with each group determined to articulate and defend their viewpoints.

Throughout history, we can discern a recurring pattern wherein religious authorities initially resisted the advancement of new scientific discoveries. One notable example of this tension is found in the reluctance of Charles Darwin to publish his theory of Evolution by Natural Selection. Darwin was acutely aware that his revolutionary ideas could be perceived as a direct challenge to religious doctrines, particularly the biblical account of creation. He recognized that his theory had the potential to disrupt the deeply ingrained religious and scientific beliefs of his era. He anticipated formidable opposition and feared harsh criticism from both religious authorities and the scientific community. He harbored fears of suffering a fate similar to that of Giordano Bruno and Galileo Galilei.

Giordano Bruno, a philosopher, cosmologist, and mathematician, met a gruesome end in the year 1600, as he was condemned to death by burning at the stake. His tragic demise was the result of his unorthodox views about the universe, which ran afoul of prevailing religious doctrines. Similarly, Galileo Galilei, a philosopher, astronomer, and mathematician, faced a harsh verdict in 1633. He was sentenced to a lifetime of house arrest, which lasted until his passing in 1642. His transgression lay in his revolutionary discovery that the Earth orbited the Sun—a proposition that clashed with established beliefs derived from religious scriptures.

A parallel scenario unfolded when the theory of Human Evolution was initially introduced. Some religious authorities and theologians, especially those hailing from more conservative and literalist traditions, vehemently resisted and rejected the theory of evolution.

They perceived it as fundamentally incompatible with their understanding of a literal interpretation of sacred texts, such as the Bible's accounts of creation.

There exists a widespread misconception that embracing the theory of human evolution necessitates the rejection of a higher power or the denial of God's existence. Many individuals perceive the acceptance of human evolution as a stepping stone towards atheism or agnosticism, and some even believe that it prompts individuals to cast aside their deeply held religious convictions, deeming them incompatible with the evolutionary framework.

Renowned biologist and vocal advocate of atheism, Richard Dawkins, stands as a prominent supporter of the theory of evolution. In his perspective, the revelation of evolution has afforded him a sense of intellectual fulfilment, one he describes as being an "intellectually fulfilled atheist." For Dawkins, Charles Darwin's groundbreaking theory of natural selection provides a compelling answer to the age-old question of human origins, obviating the need for the intervention of a divine entity. In his worldview, the intricate web of life's diversity, as unveiled by evolution, operates independently of any divine assistance.

A prevalent idea suggests an inseparable connection between Darwinism and atheism, often portraying them as intertwined aspects of a single concept. This raises the question: Is this perception firmly grounded in undeniable truth, or has it evolved into a belief of its own over time?

Parallel to the wide spectrum of religious beliefs, the intertwining of Darwinism and atheism has become firmly embedded in public discourse. Yet, it's worth contemplating whether the reinforcement of this belief is, to some extent, a result of insufficient, intellectual engagement between theologians and those embracing the theory of evolution.

Perhaps the key to a more nuanced understanding lies in fostering open and rigorous dialogues that transcend the confines of preconceived notions, allowing for a deeper exploration of the intricate relationship between evolutionary science and spiritual beliefs.

The response of Muslims to the theory of evolution, akin to that of many other religious communities, has been characterized by diversity and multiplicity of viewpoints. There is no single, monolithic stance among Muslims regarding this scientific theory; opinions span a wide spectrum. Some conservative or traditionalist Muslim scholars and communities have steadfastly rejected the theory of evolution. They often assert that it challenges a literal interpretation of the Quran's narrative of creation and may advocate for a more literal understanding of a six-day creation narrative. This perspective aligns with the belief that the Earth is stationary—a viewpoint articulated by a scholar well over a century ago, a viewpoint steadfastly upheld by contemporary adherents.

Conversely, other Muslim scholars have embraced the notion of reconciling their religious beliefs with the theory of evolution. They argue that evolution represents a natural process that was set in motion by God, and they contend that it need not be inherently incompatible with the Quran or Islamic teachings. These scholars perceive evolution as a means through which God initiated and diversified life on Earth.

In summary, the interplay between scientific discoveries, religious beliefs, and the evolution of thought continues to shape the discourse surrounding human evolution. This dynamic interaction has unfolded over centuries, resulting in diverse perspectives and ongoing discussions among scholars, theologians, and the faithful.

PART II: Research

It is crucial to underscore that the research conducted on this subject does not aim to confront or dismiss the existing beliefs of any religious group or sect. Rather, its purpose is to foster a deeper comprehension of the Quranic perspective.

The predominant belief among Muslims, often labeled as creationism, asserts that the genesis of human life traces back to the direct creation of Adam by God. In accordance with this perspective, Eve is believed to have been created from Adam, marking the initiation of the human reproductive process. Central to this belief is the notion that both Adam and Eve were created directly by the divine, exempt from the involvement of biological parents, thereby rendering the inquiry into human evolution inconsequential.

It is notable that a prevailing perspective among many Muslims views the acceptance of human evolution as Haraam (forbidden) and Kufr (a rejection of God's divine order). This viewpoint, deeply entrenched in the tradition of Islamic scholarship, does not find its origins in religious texts such as the Quran and Hadith (the sayings of the Prophet Muhammad). Rather, it has evolved over time through the elaboration and reinforcement by numerous scholars. The persistence of this belief can be partly attributed to limited exposure to scientific knowledge or perhaps a reluctance to delve into the complexities of scientific inquiry. Additionally, it may stem from the prevailing tendency to interpret the Quran in alignment with established perspectives rather than embarking on an exploratory journey within its verses, with the intention of guiding common understanding toward a more enlightened path.

In the forthcoming chapter, titled "Extraction of Verses," we shall delve into the examination of pertinent Quranic verses and their associated common interpretations. This exploration aims to shed light on why the prevailing belief in creationism, as outlined above, has persevered over time and continues to be an integral aspect of the faith of many Muslims. In the chapters subsequent to the chapter Four:

"Extraction of Verses," we aim to present interpretations without preconceived outcomes or reliance on commonly accepted narratives.

Chapter 3 Verse 7 of the Quran states, *"It is He Who has revealed the Book to you. Some of its verses are clear and lucid, and these are the core of the Book. Others are ambiguous. Those in whose hearts there is perversity, always go about the part which is ambiguous, seeking mischief and seeking to arrive at its meaning arbitrarily, although none knows their true meaning except God. On the contrary, those firmly rooted in knowledge say: 'We believe in it; it is all from our Lord alone.' No one derives true admonition from anything except the men of understanding."*

Within the domain of literature and language, the Quran presents distinct classifications of its verses. It identifies them as **Muh'kamaat**, denoting the clear and unambiguous verses, and **Muta'shabihaat**, signifying those that are allegorical or metaphorical in nature. The Muh'kamaat verses primarily provide explicit directives, leaving no room for alternative interpretations. They straightforwardly convey commands to either engage in specific actions or refrain from certain behaviors.

On the other hand, the Muta'shabihaat verses, by their very nature, are open to a range of interpretations. They employ metaphor and symbolism, and their meaning may vary depending on the level of knowledge and insight of the interpreter. These verses invite deeper contemplation and scholarly study to unravel their intended significance, allowing for diverse understandings that align with the interpreter's comprehension and expertise.

The core of this book's investigation revolves around the Muta'shabihaat verses of the Quran. Consequently, if individuals hold dissimilar interpretations or harbor distinct opinions regarding the research's ultimate findings, such diversity aligns with the foundational principles of Islamic thought. The Islamic context inherently accommodates varied understandings, encouraging a respectful acknowledgement of differing viewpoints.

Moreover, the outcomes derived from this research should be perceived as informed perspectives. Even if not universally accepted, these conclusions merit considerate respect within the Islamic framework. In the spirit of Islamic discourse, there exists a commitment to constructive dialogue and a recognition of diverse viewpoints, especially when grappling with the nuanced and metaphorical dimensions of the Quranic text.

The Quran urges individuals to embark on journeys, both physical and intellectual, to gain a deeper understanding of the origins and purpose of the natural world. This call to exploration and reflection is an invitation to connect with the profound wisdom and purpose embedded in the creation of the universe.

Chapter 29 Verse 20 clearly states, *Say, "Travel in the land and see how He originated creation, and then God will bring forth (resurrect) the creation of the Hereafter (i.e. resurrection after death). Verily, God is able to do all things."*

Chapter Four: Extraction of Verses

Within the sacred text of the Quran, there lies a profound concept that sets it apart—the verses, referred to as "Ayaah", serve as poignant reminders of divine messages. Ayaah, translated literally as "signs", encapsulate the multifaceted nature of the Quranic revelations. Much like the verses found within the Bible, these Ayaah encompass revelations that guide, enlighten, and inspire.

The process of extracting verses relevant to a research work is comparable to the careful curation of a detailed tapestry. It entails the deliberate choice of verses that hold profound significance for the topic under scrutiny. These chosen verses, much like treasures within a vault, offer deep insights and wisdom, drawing from the timeless reservoir of Quranic knowledge.

The purpose of extracting these verses is to seamlessly integrate them into subsequent chapters of our work, providing only the translation. This approach aims to aid the reader in understanding the common interpretation and public understanding of the Quranic text. Additionally, for those who wish to conduct further research or verify references independently, this format offers convenience. Among the various translations available from renowned scholars, the translation utilized here is from "Sahih International," a project of the Saheeh International organization. This organization is committed to producing accurate and reliable translations of the Quran.

Group 1: Origin of Life

Chapter 04, An-Nisa, Verse 01:

يَا أَيُّهَا النَّاسُ اتَّقُوا رَبَّكُمُ الَّذِيْ خَلَقَكُمْ مِنْ نَّفْسٍ وَّاحِدَةٍ وَّ خَلَقَ مِنْهَا زَوْجَهَا وَ بَثَّ مِنْهُمَا رِجَالًا كَثِيْرًا وَّ نِسَاءً ۚ وَ اتَّقُوا اللهَ الَّذِيْ تَسَاءَلُوْنَ بِهِ وَ الْأَرْحَامَ ۚ إِنَّ اللهَ كَانَ عَلَيْكُمْ رَقِيْبًا

Transliteration: Ya ayyuhan-nasu-ttaqoo rabbakumu allathee khalaqakum min nafsin wahidati-wakhalaqa minha zawjaha wabaththa minhuma rijalan katheeran wanisaa wa-ttaqoo-llaha

The Quranic narrative on Human Evolution

allathee tasaaloona bihi wal-arhama inna Allaha kana A'laykum raqeeban.

Translation: O mankind, fear your Lord, who created you from one soul and created from it its mate and dispersed from both of them many men and women. And fear Allah, through whom you ask one another, and the wombs [that bore you]. Indeed, Allah is ever an Observer over you.

Common Interpretation: This verse emphasizes the unity of the human race, originating from one soul (Adam) and couple (Eve) created from him, and highlights the importance of fearing God and maintaining family ties and relationships.

Chapter 6, Al-Anaam, Verse 98:

وَهُوَ الَّذِيْ اَنْشَاَكُمْ مِّنْ نَّفْسٍ وَّاحِدَةٍ فَمُسْتَقَرٌّ وَّ مُسْتَوْدَعٌ ۚ قَدْ فَصَّلْنَا الْآيٰتِ لِقَوْمٍ يَّفْقَهُوْنَ

Transliteration: Wahuwa allathee anshaakum min nafsin wahidatin famustaqarrun wamustawdaAAun qad fassalna alayati liqawmin yafqahoon.

Translation: And it is He who produced you from one soul and [gave you] a place of dwelling and of storage. We have detailed the signs for a people who understand.

Common Interpretation: This verse emphasizes the origin of humanity from a single soul (Adam) and the provision of a place for dwelling and sustenance. It underscores the signs and revelations for those who seek understanding.

Chapter 7, Al-Araaf, Verse 189:

هُوَالَّذِيْ خَلَقَكُمْ مِّنْ نَّفْسٍ وَّاحِدَةٍ وَّ جَعَلَ مِنْهَا زَوْجَهَا لِيَسْكُنَ اِلَيْهَا ۚ فَلَمَّا تَغَشّٰهَا حَمَلَتْ حَمْلًا خَفِيْفًا فَمَرَّتْ بِهٖ ۚ فَلَمَّآ اَثْقَلَتْ دَّعَوَا اللّٰهَ رَبَّهُمَا لَئِنْ اٰتَيْتَنَا صَالِحًا لَّنَكُوْنَنَّ مِنَ الشّٰكِرِيْنَ

60

The Quranic narrative on Human Evolution

Transliteration: Huwa alladhī khalaqakum min nafsin wāḥidatin waja'ala minhā zawjaha liyaskuna ilayhā falammā taghshāhā ḥamalat ḥamlan khafīfan famarrat bihi fa-lammā athqalat da'awā Allāha rabbahumā la'in ātaytanā ṣāliḥan lanakūnnanna mina al-shshākirīn

Translation: It is He who created you from one soul and created from it its couple that he might dwell in security with her. And when he covers her, she carries a light burden and continues therein. But when it becomes heavy, they both invoke Allah, their Lord, "If You should give us a good [child], we will surely be among the grateful."

Common Interpretation: This verse highlights the creation of human beings from a single soul (Adam) and the creation of mates from that soul (Eve) to provide companionship and comfort. When the burden of pregnancy becomes heavy, they both call upon Allah, seeking His blessings and expressing gratitude for a good outcome.

Chapter 19, Maryam, Verse 9:

قَالَ كَذَلِكَ قَالَ رَبُّكَ هُوَ عَلَيَّ هَيِّنٌ وَ قَدْ خَلَقْتُكَ مِنْ قَبْلُ وَ لَمْ تَكُ شَيْئًا

Transliteration: Qala kadalika. Qala rabbuka huwa alayya hayyinun wa qad khalaqtuka min qablu wa lam taku shay'an.

Translation: [An angel] said, Thus [it will be]; your Lord says, It is easy for Me, for I created you before, while you were nothing.

Common Interpretation:

In this verse, the angel responds to Maryam (Mary) with the message from her Lord, stating that the miraculous birth of Isa (Jesus) is something decreed and easy for Allah. The verse emphasizes the divine will and power in bringing about extraordinary events, and it foreshadows the birth of Prophet Isa as a miraculous sign and a mercy from Allah.

Chapter 19, Maryam, Verse 67:

أَوَ لَا يَذْكُرُ الْإِنْسَانُ أَنَّا خَلَقْنَاهُ مِنْ قَبْلُ وَ لَمْ يَكُ شَيْئًا

Transliteration: Awa lā yadhkurul-insānu annā khalaqnāhu min qablu wa lam yaku shay'an

Translation: Does man not remember that We created him before, while he was nothing?

Common Interpretation: In this verse, Allah questions whether humans do not reflect upon the fact that Allah created them when they were nothing before. It serves as a reminder of the origin of mankind from a state of non-existence and highlights the creative power of Allah. The verse encourages contemplation on the miraculous process of human creation and the divine wisdom behind it.

Chapter 39, Az-Zumr, Verse 06:

خَلَقَكُمْ مِنْ نَفْسٍ وَّاحِدَةٍ ثُمَّ جَعَلَ مِنْهَا زَوْجَهَا وَ أَنْزَلَ لَكُمْ مِنَ الْأَنْعَامِ ثَمَٰنِيَةَ أَزْوَاجٍ ۚ يَخْلُقُكُمْ فِىْ بُطُوْنِ أُمَّهَٰتِكُمْ خَلْقًا مِّنْ بَعْدِ خَلْقٍ فِىْ ظُلُمَٰتٍ ثَلٰثٍ ۚ ذَٰلِكُمُ اللّٰهُ رَبُّكُمْ لَهُ الْمُلْكُ ۚ لَآ اِلَٰهَ اِلَّا هُوَ فَاَنّٰى تُصْرَفُوْنَ

Transliteration: Khalaqakum min nafsin wāḥidatin thumma ja ala minhā zawjaha wa anzala lakum mina al-an āmi thamāniyata azwājin, yakhluqukum fī buṭūni ummahātikum khalaqan min ba'da khalaqin fī ẓulumātin thalāthin. Dhālikumu Allāhu rabbukum, lahu al-mulku, lā ilāha illā huwa. Fa-annā tuṣrafun?

Translation: He created you from one soul; then He made from it its couple (spouse), and He produced for you from the grazing livestock eight mates. He creates you in the wombs of your mothers, creation after creation, within three darknesses. That is Allah, your Lord; to Him belongs dominion. There is no deity except Him, so how are you averted?

Common Interpretation: This verse is a reflection on the creation of humans and animals, emphasizing the oneness of God as the Creator and the importance of recognizing His dominion and uniqueness. It also alludes to the development of the fetus in the womb as a sign of God's creative power.

Group 2: Creation with Mud

Chapter 06, Al-Anaam, Verse 02:

هُوَ الَّذِىْ خَلَقَكُمْ مِّنْ طِيْنٍ ثُمَّ قَضٰۤى اَجَلًا ۚ وَ اَجَلٌ مُّسَمًّى عِنْدَهٗ ثُمَّ اَنْتُمْ تَمْتَرُوْنَ

Transliteration: Huwa allathee khalaqakum min teenin thumma qada ajalan waajalun musamman AAindahu thumma antum tamtaroon

Translation: It is He (God) who created you from clay and then decreed a term and a specified time [known] to Him; then [still] you are in dispute.

Common Interpretation: This verse acknowledges that God is the Creator who fashioned human beings from clay, and He has ordained a specific duration for everyone's life. The verse also highlights that this predetermined lifespan is known to God alone.

Chapter 07, Al-Araaf, Verse 12:

قَالَ مَا مَنَعَكَ اَلَّا تَسْجُدَ اِذْ اَمَرْتُكَ ۚ قَالَ اَنَا خَيْرٌ مِّنْهُ ۚ خَلَقْتَنِىْ مِنْ نَّارٍ وَّ خَلَقْتَهٗ مِنْ طِيْنٍ

Transliteration: Qala ma manAAaka alla tasjuda ith amartuka qala ana khayrun minhu khalaqtanee min narrin wakhalaqtahu min teen

Translation: [Allah] said, "What prevented you from prostrating when I commanded you?" [Satan] said, "I am better than him. You created me from fire and created him from clay [i.e., earth]."

Common Interpretation: In this verse, God questions Satan as to why he did not prostrate when commanded to do so, and Satan responds with arrogance, claiming to be superior to Adam. Satan

The Quranic narrative on Human Evolution

justifies his refusal to prostrate to Adam by stating that he was created from fire, while Adam was created from clay.

Chapter 15, Al-Hijr, Verse 26:

وَ لَقَدْ خَلَقْنَا الْإِنْسَانَ مِنْ صَلْصَالٍ مِنْ حَمَإٍ مَّسْنُونٍ

Transliteration: Wa laqad khalaqnal-insana min sal salin min hama'in masnoon

Translation: And We certainly created man out of clay from an altered black mud.

Common Interpretation: In this verse, it is stated that God created human beings from clay, specifically from an altered, black, and moldable mud. This verse emphasizes the origin of human creation and the humble substance from which humanity was formed. It serves as a reminder of the divine power and creativity in shaping human life from simple earthly elements.

Chapter 15, Al-Hijr, Verse 33:

قَالَ لَمْ أَكُنْ لِأَسْجُدَ لِبَشَرٍ خَلَقْتَهُ مِنْ صَلْصَالٍ مِنْ حَمَإٍ مَّسْنُونٍ

Transliteration: Qāla lam akun li-asjuda li-basharin khalaqtahu min ṣalṣālin min ḥamā'in masnūn

Translation: He said, "Never would I prostrate to a human whom You created out of clay from an altered black mud."

Common Interpretation: In this verse, it describes a moment when Allah (God) created Adam (peace be upon him) from clay, a special type of clay that was altered and fashioned by Him. Iblis refused the order from God to bow down to Adam, citing his belief that he was superior because he was created from fire while Adam was created from clay.

Chapter 17, Bani-Israeel, Verse 61:

وَ إِذْ قُلْنَا لِلْمَلَٰٓئِكَةِ اسْجُدُوا لِآدَمَ فَسَجَدُوٓا إِلَّآ إِبْلِيسَ قَالَ ءَأَسْجُدُ لِمَنْ خَلَقْتَ طِينًا

Transliteration: Wa idh qulna lil-malā'ikati isjudū li-ādama fasajadū illā iblīsa qāla aasjudu liman khalaqta ṭīnā

Translation: And [mention] when We said to the angels, prostrate to Adam, and they prostrated, except for Iblis. He said, Should I prostrate to one You created from clay?

Common Interpretation: This verse recounts the moment when Allah commanded the angels and djinns to prostrate before Adam as a sign of respect and honor. Iblis (Satan) refused the order from God to bow down to Adam, his refusal was rooted in his arrogance and pride, as he considered himself superior to Adam, believing that he was created from fire while Adam was created from clay.

Chapter 23, Al-Mominoon, Verse 12:

وَ لَقَدْ خَلَقْنَا الْإِنْسَانَ مِنْ سُلَٰلَةٍ مِنْ طِينٍ

Transliteration: Wa laqad khalaqna al-insana min sulālatin min ṭīn.

Translation: And certainly did We create man from an extract of clay.

Common Interpretation: This verse succinctly mentions the creation of human beings. It emphasizes that humans (more specifically Adam) were created from clay.

Chapter 32, As-Sajdah, Verse 07:

الَّذِىٓ أَحْسَنَ كُلَّ شَىْءٍ خَلَقَهُۥ وَ بَدَأَ خَلْقَ الْإِنْسَانِ مِنْ طِينٍ

Transliteration: Alladhī aḥsana kulla shayin khalaqahu wa-badā khalka al-insāni min ṭīn.

Translation: Who perfected everything which He created and began the creation of man from clay.

Common Interpretation: This verse highlights the perfection of God's creation. It acknowledges that God is the One who has perfected everything He created. The verse also emphasizes the initial stage of human creation, where Allah began the creation of mankind (by creating Adam) from clay or dust.

Chapter 37, As-Saffaat, Verse 11:

فَاسْتَفْتِهِمْ أَهُمْ أَشَدُّ خَلْقًا أَم مَّنْ خَلَقْنَا ۚ إِنَّا خَلَقْنَٰهُم مِّن طِينٍ لَّازِبٍ

Transliteration: Fas-taftihim ahum ashaddu khalqan am man khalaqnā? Innā khalaqnāhum min ṭīnin lazib.

Translation: Then inquire of them, [O Muhammad], "Are they a stronger [or more difficult] creation or those [others] We have created?" Indeed, We created them [i.e., men] from sticky clay.

Common Interpretation: This verse conveys a message emphasizing the rhetorical question about the strength and creation of human beings and the recognition that all humans, including the disbelievers, were created from clay.

Chapter 38, Suaad, Verse 76:

قَالَ أَنَا خَيْرٌ مِّنْهُ خَلَقْتَنِى مِن نَّارٍ وَخَلَقْتَهُ مِن طِينٍ

Transliteration: Qāla anā khayrun minhu khalaqtanī min nārin wa khalaqtahu min ṭīn.

Translation: He said, "I am better than him; You created me from fire, and You created him from clay."

Common Interpretation: In this verse, Iblis (Satan) expresses his arrogance and defiance by claiming that he is superior to Adam (peace be upon him) because he was created from fire, while Adam was created from clay. This statement reflects Iblis's pride and disobedience to Allah's command, leading to his expulsion from Paradise.

Chapter 55, Ar-Rahmaan, Verse 14:

خَلَقَ الْإِنْسَانَ مِنْ صَلْصَالٍ كَالْفَخَّارِ

Transliteration: Khalaqal-insāna min ṣalsālin kaalfakhkār.

Translation: He created man from clay like [that of] pottery.

Common Interpretation: This verse highlights the creation of humans from clay, emphasizing the humble origin of human beings as being fashioned from earthly materials. It serves as a reminder of Allah's creative power and His ability to bring life into existence from the simplest elements.

Chapter 71, Nuh, Verse 17:

وَ اللهُ أَنْبَتَكُمْ مِّنَ الْأَرْضِ نَبَاتًا

Transliteration: Wa Allāhu anbatukum min al-arḍi nabātā

Translation: And Allah has caused you to grow from the earth a [progressive] growth.

Common Interpretation: This verse is a reminder of the creation of human beings. It speaks of how God created human beings from the earth, referring to the initial creation of Adam.

Group 3: Creation with Water

Chapter 21, Al-Ambiya, Verse 30

اَوَ لَمْ يَرَ الَّذِيْنَ كَفَرُوْٓا اَنَّ السَّمٰوٰتِ وَ الْاَرْضَ كَانَتَا رَتْقًا فَفَتَقْنٰهُمَاؕ وَ جَعَلْنَا مِنَ الْمَآءِ كُلَّ شَيْءٍ حَيٍّؕ اَفَلَا يُؤْمِنُوْنَ

Transliteration: Awalam yara alladhīna kafarū anna as-samāwāti wal-arḍa kānatā ratqan fa-fataqnāhumā, wa ja'alnā minal-mā'i kulla shayin ḥayyi. Afalā yu'minūn?

Translation: Have those who disbelieved not considered that the heavens and the earth were a joined entity, and then We separated them and made from water every living thing? Then will they not believe?

Common Interpretation: In this verse, God challenges those who disbelieve to reflect upon the creation of the heavens and the earth. It mentions that the heavens and the earth were originally one closed or joined entity (ratqan), and Allah opened and separated them. Furthermore, it highlights that Allah created every living thing from water.

Chapter 24, An-Noor, Verse 45

وَ اللّٰهُ خَلَقَ كُلَّ دَآبَّةٍ مِّنْ مَّآءٍۚ فَمِنْهُمْ مَّنْ يَّمْشِيْ عَلٰى بَطْنِهٖۚ وَ مِنْهُمْ مَّنْ يَّمْشِيْ عَلٰى رِجْلَيْنِۚ وَ مِنْهُمْ مَّنْ يَّمْشِيْ عَلٰٓى اَرْبَعٍؕ يَخْلُقُ اللّٰهُ مَا يَشَآءُؕ اِنَّ اللّٰهَ عَلٰى كُلِّ شَيْءٍ قَدِيْرٌ

Transliteration: Wa-Allāhu khalaqa kulla dābbatin min mā'in, faminhum man yamshi alā baṭnihī, wa-minhum man yamshi alā rijlayn, wa-minhum man yamshi alā arba'in. Yakhluqu Allāhu mā yashā'u. Innallāha alā kulli shay'in qadīr.

Translation: Allah has created every [living] creature from water. And of them are those that move on their bellies, and of them are those that walk on two legs, and of them are those that walk on four. Allah creates what He wills. Indeed, Allah is over all things competent.

The Quranic narrative on Human Evolution

Common Interpretation: This verse highlights God's creative power in forming various creatures from water. It mentions that there are creatures that move upon their bellies, some that move on two legs, and others that move on four. God's creative ability is emphasized, as He creates and designs all living beings according to His will. It underscores God's omnipotence and His control over all aspects of creation.

Chapter 25, Al-Furqaan, Verse 54

وَ هُوَ الَّذِيْ خَلَقَ مِنَ الْمَآءِ بَشَرًا فَجَعَلَهُ نَسَبًا وَّ صِهْرًا ۗ وَ كَانَ رَبُّكَ قَدِيْرًا

Transliteration: Wa huwa alladhī khalaqa mina al-mā'i basharan faja'alahū nasaban wa ṣihrā, wa kāna rabbuka qadīrā.

Translation: And it is He who has created from water a human being and made him [a relative by] lineage and marriage. And ever is your Lord competent [concerning creation].

Common Interpretation: This verse from Surah Al-Furqan acknowledges God's creative power in forming human beings from water. It also emphasizes two significant aspects of lineage and marriage that God created humans and established family relationships through lineage and marriage. This highlights the importance of family and social connections in human society.

Group 4: Stages of Creation
Chapter 18, Al-Kahf, Verse 37:

قَالَ لَهُ صَاحِبُهُ وَ هُوَ يُحَاوِرُهُ اَكَفَرْتَ بِالَّذِيْ خَلَقَكَ مِنْ تُرَابٍ ثُمَّ مِنْ نُطْفَةٍ ثُمَّ سَوّٰىكَ رَجُلًا

Transliteration: Qāla lahu ṣāḥibu-hu wa-huwa yuḥāwiruhu akafarta billadhī khalaqaka min turābin thumma min nuṭfatin thumma sawwāka rajulan.

Translation: His companion said to him while he was conversing with him, "Have you disbelieved in He who created you from dust and then from a sperm-drop and then proportioned you [as] a man?"

Common Interpretation: The verse recounts a conversation between two individuals, one of whom questions the disbelief of his companion. The companion is reminded of the miraculous process of his creation, from being created from dust, then from a sperm-drop, and then being fashioned into a human being by God.

Chapter 22, Al-Hajj, Verse 05:

يَا أَيُّهَا النَّاسُ إِنْ كُنْتُمْ فِى رَيْبٍ مِنَ الْبَعْثِ فَإِنَّا خَلَقْنَاكُمْ مِنْ تُرَابٍ ثُمَّ مِنْ نُطْفَةٍ ثُمَّ مِنْ عَلَقَةٍ ثُمَّ مِنْ مُضْغَةٍ مُخَلَّقَةٍ وَ غَيْرِ مُخَلَّقَةٍ لِنُبَيِّنَ لَكُمْ وَ نُقِرُّ فِى الْأَرْحَامِ مَا نَشَاءُ إِلَى أَجَلٍ مُسَمَّى ثُمَّ نُخْرِجُكُمْ طِفْلًا ثُمَّ لِتَبْلُغُوا أَشُدَّكُمْ وَ مِنْكُمْ مَنْ يُتَوَفَّى وَ مِنْكُمْ مَنْ يُرَدُّ إِلَى أَرْذَلِ الْعُمُرِ لِكَيْلَا يَعْلَمَ مِنْ بَعْدِ عِلْمٍ شَيْئًا وَ تَرَى الْأَرْضَ هَامِدَةً فَإِذَا أَنْزَلْنَا عَلَيْهَا الْمَاءَ اهْتَزَّتْ وَ رَبَتْ وَ أَنْبَتَتْ مِنْ كُلِّ زَوْجٍ بَهِيجٍ

Transliteration: Yā ayyuhā an-nāsu in kuntum fī raibin min al-ba'thi fa-innā khalaqnākum min turābin thumma min nuṭfatin thumma min alaqatin thumma min muḍghatin mukhallaqatin wa-ghayri mukhallaqatin li-nubayyina lakum. Wa nuqirru fī al-arḥāmi mā nashā'u ilā ajalin musammā, thumma nukhriju-kum ṭiflan thumma li-tablughū ašuddakum. Wa minkum man yutawaffā wa minkum man yuraddu ilā arẓali al-umuri li-kaylā ya'lama mim ba'da ilmin shay'an. Wa tarā al-arḍa hāmidatan fa-idhā anzalnā alayhā al-mā'a ihtazzat wa-rabat wa-anbatat min kulli zawjim baḥījin.

Translation: O people, if you should be in doubt about the Resurrection, then [consider that] indeed, We created you from dust, then from a sperm-drop, then from a clinging clot, and then from a lump of flesh, formed and unformed – that We may show you. And We settle in the wombs whom We will for a specified term, then We bring you out as a child, and then [We develop you] that you may reach your [time of] maturity. And among you is he who is taken in [early] death, and among you is he who is returned to the most decrepit [old] age so that he knows, after [once having] knowledge, nothing.

70

The Quranic narrative on Human Evolution

And you see the earth barren, but when We send down upon it rain, it quivers and swells and grows [something] of every beautiful kind.

Common Interpretation: This verse highlights the process of human creation, development, and aging. It also serves as a reminder of Allah's power to create and resurrect human beings, even if they have doubts about the concept of resurrection. It emphasizes the gradual stages of human development, from a mere sperm-drop to a fully developed individual, and how human life is ultimately in Allah's control.

Chapter 35, Fatir, Verse 11:

وَ اللهُ خَلَقَكُمْ مِنْ تُرَابٍ ثُمَّ مِنْ نُطْفَةٍ ثُمَّ جَعَلَكُمْ أَزْوَاجًا ۚ وَ مَا تَحْمِلُ مِنْ أُنْثَى وَ لَا تَضَعُ إِلَّا بِعِلْمِهِ ۚ وَ مَا يُعَمَّرُ مِنْ مُعَمَّرٍ وَ لَا يُنْقَصُ مِنْ عُمُرِهِ إِلَّا فِيْ كِتٰبٍ ۚ إِنَّ ذٰلِكَ عَلَى اللهِ يَسِيْرٌ

Transliteration: Wallāhu khalaqakum min turābin thumma min nuṭfatin thumma ja'alakum azwājā. Wa mā taḥmilu min unthā walā taḍa'u illā bi-ilmihi. Wa mā yu'ammaru min mu'ammarin walā yunqaṣu min umurihi illā fī kitābin. Inna dhālika alāllāhi yasīr.

Translation: And Allah created you from dust, then from a sperm-drop; then He made you mates. And no female conceives, nor does she give birth except with His knowledge. And no aged person is granted [additional] life nor is his lifespan lessened but that it is in a register. Indeed, that for Allah is easy.

Common Interpretation: This verse discusses the process of human creation and the divine knowledge and control over life and death. It highlights God created human beings from dust and then from a sperm-drop, God has ordained the concept of mates and procreation, the conception and birth of every individual are known only to Allah, and it happens with His knowledge and permission and that the length of an individual's life is determined by Allah and is recorded in a register (book of decrees).

The Quranic narrative on Human Evolution

Group 5: Bashar (Pre-human being)
Chapter 15, Al-Hijr, Verse 28:

وَ اِذْ قَالَ رَبُّكَ لِلْمَلٰٓئِكَةِ اِنِّىْ خَالِقٌۢ بَشَرًا مِّنْ صَلْصَالٍ مِّنْ حَمَاٍ مَّسْنُوْنٍ

Transliteration: Wa ith qāla rabbuka lilmalā'ikati innī khāliqun basharan min ṣalṣālin min ḥama'in masnūn

Translation: And [mention, O Muhammad], when your Lord said to the angels, "I will create a human being out of clay from an altered black mud."

Common Interpretation: This verse describes the moment when Allah informed the angels of His intention to create a human being, Adam, from clay or molded mud. It highlights the creative power of Allah in shaping humanity from humble materials.

Chapter 15, Al-Hijr, Verse 33:

قَالَ لَمْ اَكُنْ لِّاَسْجُدَ لِبَشَرٍ خَلَقْتَهٗ مِنْ صَلْصَالٍ مِّنْ حَمَاٍ مَّسْنُوْنٍ

Transliteration: Qāla lam akun li'asjuda li-basharin khalaqtahu min ṣalṣālin min ḥamā in masnūn

Translation: He said, "Never would I prostrate to a human whom You created out of clay from an altered black mud."

Common Interpretation: In this verse, Iblis (Satan) is responding to Allah's command to prostrate to Adam. Iblis refuses, objecting that he will not bow to a human created from clay, specifically from an altered black mud. This verse is part of the narrative about Iblis' disobedience and his refusal to acknowledge the superiority of Adam as the first human.

Chapter 25, Al-Furqaan, Verse 54:

وَ هُوَ الَّذِىْ خَلَقَ مِنَ الْمَآءِ بَشَرًا فَجَعَلَهٗ نَسَبًا وَّ صِهْرًا ؕ وَ كَانَ رَبُّكَ قَدِيْرًا

The Quranic narrative on Human Evolution

Transliteration: Wa huwa alladhī khalaqa min al-mā'i basharan faja'alahū nasabāw wa ṣiḥrā. Wa kāna rabbuka qadīrā.

Translation: And it is He who has created from water a human being and made him [a relative by] lineage and marriage. And ever is your Lord competent [concerning creation].

Common Interpretation: This verse highlights God's role in the creation of human beings from water and the establishment of family and marital relationships among them. It emphasizes God's ability and power in shaping and governing His creation.

Chapter 30, Ar-Room, Verse 20:

وَ مِنْ اٰيٰتِهٖٓ اَنْ خَلَقَكُمْ مِنْ تُرَابٍ ثُمَّ اِذَآ اَنْتُمْ بَشَرٌ تَنْتَشِرُوْنَ

Transliteration: Wa min ayatihi an khalaqakum min turabin thumma idha antum basharun tantashirun

Translation: And of His signs is that He created you from dust; then, suddenly you were human beings dispersing [throughout the earth].

Common Interpretation: This verse highlights one of the signs of Allah's power and creation, emphasizing that He created human beings from dust and, in a miraculous process, they became human beings scattered across the earth. It underscores the concept of the origin of humankind from humble beginnings and their subsequent proliferation across the world.

Chapter 38, Saad, Verse 71:

اِذْ قَالَ رَبُّكَ لِلْمَلٰٓئِكَةِ اِنِّىْ خَالِقٌۢ بَشَرًا مِنْ طِيْنٍ

Transliteration: Iz qāla rabbuka lil-malā'ikati innī khāliqum basharan min ṭīn

Translation: [So mention] when your Lord said to the angels, "Indeed, I am going to create a human being from clay."

Common Interpretation: This verse narrates the moment when God informed the angels about His act of creating a human being from clay. It is part of the story of the creation of Adam and emphasizes the divine origin of humanity, highlighting that human beings are created from humble materials.

Chapter 76, Ad-Dahar, Verse 01:

بَلْ اَتٰى عَلَى الْاِنْسَانِ حِيْنٌ مِّنَ الدَّهْرِ لَمْ يَكُنْ شَيْئًا مَّذْكُوْرًا

Transliteration: Hal atā alā al-insāni ḥīnun mina ad-dahri lam yakun shay'an maẓkūrā.

Translation: Has there [not] come upon man a period of time when he was not a thing [even] mentioned?

Common Interpretation: This verse in the Quran refers to a period in the existence of humans when they were not even worth mentioning. The verse serves to remind humans of their humble origins and the contrast between their limited knowledge and God's all-encompassing knowledge.

Group 6: First Human Being

Chapter 2, Al-Baqarah, Verse 30:

وَ اِذْ قَالَ رَبُّكَ لِلْمَلٰٓئِكَةِ اِنِّىْ جَاعِلٌ فِى الْاَرْضِ خَلِيْفَةً ۗ قَالُوْۤا اَتَجْعَلُ فِيْهَا مَنْ يُّفْسِدُ فِيْهَا وَ يَسْفِكُ الدِّمَاۤءَ ۚ وَ نَحْنُ نُسَبِّحُ بِحَمْدِكَ وَ نُقَدِّسُ لَكَ ۗ قَالَ اِنِّىْۤ اَعْلَمُ مَا لَا تَعْلَمُوْنَ

Transliteration: Wa iz qaala rabbuka lil malaaa'ikati innee jaa'ilun fil ardi khaleefatan qaalooo ataj'alu feehaa mai yufsidu feehaa wa yasfikud dimaaa'a wa nahnu nusabbihu bihamdika wa nuqaddisu laka qaala inneee a'lamu maa laa ta'lamoon

Translation: And [mention, O Muhammad], when your Lord said to the angels, "Indeed, I will make upon the earth a successive authority." They said, "Will You place upon it one who causes corruption therein and sheds blood, while we exalt You with praise and declare Your perfection?" He [Allah] said, "Indeed, I know that which you do not know."

Common Interpretation: This verse addresses the appointment of human beings as stewards or representatives on Earth and the concerns raised by the angels regarding human behavior on the planet.

Chapter 2, Al-Baqarah, Verse 117:

بَدِيْعُ السَّمٰوٰتِ وَ الْأَرْضِ ۚ وَ إِذَا قَضٰى اَمْرًا فَاِنَّمَا يَقُوْلُ لَهٗ كُنْ فَيَكُوْنُ

Transliteration: Badī'u as-samāwāti wal-arḍi; wa-idhā qaḍā amran fa-innamā yaqūlu lahu kun fa-yakūn.

Translation: Originator of the heavens and the earth. When He decrees a matter, He only says to it, "Be," and it is.

Common Interpretation: This verse emphasizes the creative power of God as the Originator of the heavens and the earth. It highlights His ability to bring about any matter with a simple command, "Be" and it comes into existence. It underscores the divine authority and ease with which God creates and controls the universe.

Chapter 17, Bani Israeel, Verse 85:

وَ يَسْـَٔلُوْنَكَ عَنِ الرُّوْحِ ۚ قُلِ الرُّوْحُ مِنْ اَمْرِ رَبِّيْ وَ مَآ اُوْتِيْتُمْ مِّنَ الْعِلْمِ اِلَّا قَلِيْلًا

Transliteration: Wayas'alūnaka ani-rūḥi quli-rūḥu min amri rabbī wa mā ūtītum mina al-ilmi illā qalīlā

Translation: And they ask you, [O Muhammad], about the soul. Say, "The soul is of the affair of my Lord. And mankind has not been given of knowledge except a little."

Common Interpretation: In this verse, people are inquiring about the nature of the soul. The response from the Prophet Muhammad (peace be upon him) is that the soul is from the command of Allah, and human knowledge about it is limited. This verse acknowledges the limited understanding that human beings have about the nature of the soul and attributes it to the divine domain of Allah's knowledge and command.

Chapter 32, As-Sajdah, Verse 09:

ثُمَّ سَوَّىٰهُ وَ نَفَخَ فِيْهِ مِنْ رُوْحِهِ وَ جَعَلَ لَكُمُ السَّمْعَ وَ الْأَبْصَارَ وَ الْأَفْـِٕدَةَ ۚ قَلِيْلًا مَّا تَشْكُرُوْنَ

Transliteration: Thumma sawwāhu wa nafakha fīhi min rūḥihi wa ja'ala lakumu as-sam'a wal-abṣāra wal-af'idata. Qalīlan mā tashkurūn.

Translation: Then He proportioned him and breathed into him from His [created] soul and made for you hearing and vision and hearts; little are you grateful.

Common Interpretation: This verse discusses the creation of human beings. It mentions that God shaped and proportioned humans, breathed His divine spirit into them, and bestowed upon them the faculties of hearing, vision, and understanding (hearts). It is a reminder of the special status and blessings given to humanity by God. However, it also highlights the ingratitude of many people who fail to appreciate these divine gifts.

Chapter 36, Yaseen, Verse 82:

إِنَّمَآ أَمْرُهٗ إِذَآ أَرَادَ شَيْئًا أَنْ يَّقُوْلَ لَهٗ كُنْ فَيَكُوْنُ

Transliteration: Innamā amruhu idhā arāda shayan an yaqūla lahu kun fayakūn.

Translation: His command is only when He intends a thing that He says to it, "Be," and it is.

Common Interpretation: This verse highlights the absolute power and authority of God (Allah). It states that when God intends something to happen, He simply commands it to "Be," and it immediately comes into existence. It emphasizes the divine ability to create and control the universe with a mere word, demonstrating the omnipotence of God's will.

Chapter 38, Saad, Verse 71:

إِذْ قَالَ رَبُّكَ لِلْمَلَٰئِكَةِ إِنِّىْ خَالِقٌ بَشَرًا مِّنْ طِيْنٍ

Transliteration: Iz qāla rabbuka lil-malā'ikati innī khāliqum basharan min ṭīn

Translation: [So mention] when your Lord said to the angels, "Indeed, I am going to create a human being from clay."

Common Interpretation: This verse narrates the moment when God informed the angels about His act of creating a human being from clay. It is part of the story of the creation of Adam and emphasizes the divine origin of humanity, highlighting that human beings are created from humble materials.

Chapter 38, Saad, Verse 72:

فَإِذَا سَوَّيْتُهُ وَ نَفَخْتُ فِيْهِ مِنْ رُّوْحِىْ فَقَعُوْا لَهٗ سٰجِدِيْنَ

Transliteration: Fa-idhā sawwaytuhu wa nafakhtu fīhi min rūḥī faqa'ū lahu sājidīn.

Translation: So, when I have proportioned him and breathed into him of My [created] soul, then fall down to him in prostration.

Common Interpretation: This verse describes the moment when God, after shaping Adam from clay, breathed His divine spirit into him, giving life to the first human. God commanded the angels to prostrate to Adam as a sign of respect and honor for the unique creation. This event marks the distinction of the human creation and the reverence accorded to Adam by the angels, signifying his special status in God's creation.

The Quranic narrative on Human Evolution

Chapter 38, Saad, Verse 75:

قَالَ يَا إِبْلِيسُ مَا مَنَعَكَ أَن تَسْجُدَ لِمَا خَلَقْتُ بِيَدَيَّ ۖ أَسْتَكْبَرْتَ أَمْ كُنتَ مِنَ الْعَالِينَ

Transliteration: Qāla yā Iblīsu mā mana'aka an tasjuda li-mā khalaqtu bi-yadayya - Astakbarta am kuntā mina al-ālīn.

Translation: [Allah] said, "O Iblees, what prevented you from prostrating to that which I created with My hands? Were you arrogant [then], or were you [already] among the haughty?"

Common Interpretation: In this verse, God addresses Iblis (Satan) and questions why he refused to prostrate to Adam, whom God created with His two hands. God asks whether Iblis acted out of arrogance or if he had already been among the haughty.

Chapter 79, An-Naaziaat, Verses 1 – 2:

وَالنَّازِعَاتِ غَرْقًا ۙ وَالنَّاشِطَاتِ نَشْطًا

Transliteration: Wan-naz'iāti gharqā - Wan-nāshiṭāti nashṭā

Translation: By those [angels] who extract with violence - And [by] those who remove with ease.

Common Interpretation: These verses referring to the angels who extract the souls of people at the time of death. These verses collectively serve as an oath emphasizing the reality of resurrection and the afterlife.

Group 7: Adam

Chapter 2, Al-Baqarah, Verse 31:

وَعَلَّمَ آدَمَ الْأَسْمَاءَ كُلَّهَا ثُمَّ عَرَضَهُمْ عَلَى الْمَلَائِكَةِ فَقَالَ أَنبِئُونِي بِأَسْمَاءِ هَٰؤُلَاءِ إِن كُنتُمْ صَادِقِينَ

The Quranic narrative on Human Evolution

Transliteration: Wa allama Ādama al-asmā'a kullaha thumma aradhahum alā al-malā'ikati faqāla anbi'ūnī bi-asmā'i hā'ūlā'i in kuntum ṣādiqīna.

Translation: And He taught Adam the names - all of them. Then He showed them to the angels and said, "Inform Me of the names of these, if you are truthful."

Common Interpretation: In this verse, it is mentioned that God taught Adam the names of all things. Afterward, God presented these things to the angels and asked the angels to inform Him of the names of these things. This event highlights the knowledge and wisdom bestowed upon Adam and emphasizes the unique role of humans as knowledgeable beings. It is part of the story of the creation of Adam and the angels' response to God's command.

Chapter 2, Al-Baqarah, Verse 32:

قَالُوا سُبْحَٰنَكَ لَا عِلْمَ لَنَا إِلَّا مَا عَلَّمْتَنَا ۖ إِنَّكَ أَنْتَ الْعَلِيمُ الْحَكِيمُ

Transliteration: Qālū subḥānaka lā ilma lanā illā mā allamtanā, innaka antal-alīmul-ḥakīmu.

Translation: They said, "Exalted are You; we have no knowledge except what You have taught us. Indeed, it is You who is the Knowing, the Wise."

Common Interpretation: In response to God's question, the angels replied by acknowledging God's greatness and wisdom. They admitted their limited knowledge and affirmed that their knowledge is only what God has taught them. The verse emphasizes God's infinite knowledge and wisdom, contrasting the angels' limited knowledge with God's boundless understanding.

The Quranic narrative on Human Evolution

Chapter 2, Al-Baqarah, Verse 33:

قَالَ يَٰٓـَٔادَمُ أَنۢبِئۡهُم بِأَسۡمَآئِهِمۡ ۖ فَلَمَّآ أَنۢبَأَهُم بِأَسۡمَآئِهِمۡ قَالَ أَلَمۡ أَقُل لَّكُمۡ إِنِّيٓ أَعۡلَمُ غَيۡبَ ٱلسَّمَٰوَٰتِ وَٱلۡأَرۡضِ وَأَعۡلَمُ مَا تُبۡدُونَ وَمَا كُنتُمۡ تَكۡتُمُونَ

Transliteration: Qāla yā Ādama anbi'hum bi-asmā'ihim. Falammā anba'ahum bi-asmā'ihim qāla alam aqul lakum innī a'lamu ghayba as-samāwāti wal-ardhi wa a'lamu mā tubdūna wa mā kuntum taktumūna.

Translation: He said, "O Adam, inform them of their names." And when he had informed them of their names, He said, "Did I not tell you that I know the unseen [aspects] of the heavens and the earth? And I know what you reveal and what you have concealed."

Common Interpretation: In this verse, God instructs Adam to inform the angels of the names of the created things. After Adam does so, God reaffirms His knowledge of the unseen aspects of the heavens and the earth and His awareness of what is revealed and what is hidden. This emphasizes God's all-encompassing knowledge and serves as a reminder of the distinction between God's knowledge and the limited knowledge of His creations.

Chapter 2, Al-Baqarah, Verse 34:

وَإِذۡ قُلۡنَا لِلۡمَلَٰٓئِكَةِ ٱسۡجُدُواْ لِءَادَمَ فَسَجَدُوٓاْ إِلَّآ إِبۡلِيسَ أَبَىٰ وَٱسۡتَكۡبَرَ وَكَانَ مِنَ ٱلۡكَٰفِرِينَ

Transliteration: Wa-idh qulna lil-mala'ikati isjudoo li-adama fasajadoo illa ibleesa aba wastakbara wakana minal-kafireen.

Translation: And [mention] when We said to the angels, "Prostrate to Adam," and they prostrated, except for Iblis. He refused and was arrogant and became among the disbelievers.

Common Interpretation: This verse recounts the incident when Allah instructed the angels to prostrate to Adam, but Iblis (Satan) refused, displaying pride and disbelief.

The Quranic narrative on Human Evolution

Chapter 2, Al-Baqarah, Verse 35:

وَ قُلْنَا يَآدَمُ اسْكُنْ أَنْتَ وَ زَوْجُكَ الْجَنَّةَ وَ كُلَا مِنْهَا رَغَدًا حَيْثُ شِئْتُمَا ۖ وَ لَا تَقْرَبَا هَٰذِهِ الشَّجَرَةَ فَتَكُونَا مِنَ الظَّالِمِينَ

Transliteration: Wa qulna ya Adamu uskun anta wa zawjuka al-jannata wa kula minhā raghadan haythu shi'tumā walā taqrabā hādhihi ash-shajara fatakūnā mina az-zālimīn.

Translation: And We said, "O Adam, dwell, you and your wife, in Paradise and eat therefrom in [ease and] abundance from wherever you will. But do not approach this tree, lest you be among the wrongdoers."

Common Interpretation: This verse highlights the concept of obedience to divine commands and the consequences of transgressing those limits. Allah instructed Adam and his wife to dwell in paradise and enjoy its provisions freely. They were given the freedom to eat from any tree except one. They were warned not to approach that particular tree, as doing so would make them among the wrongdoers.

Chapter 2, Al-Baqarah, Verse 36:

فَأَزَلَّهُمَا الشَّيْطَانُ عَنْهَا فَأَخْرَجَهُمَا مِمَّا كَانَا فِيهِ ۖ وَ قُلْنَا اهْبِطُوا بَعْضُكُمْ لِبَعْضٍ عَدُوٌّ ۖ وَ لَكُمْ فِى الْأَرْضِ مُسْتَقَرٌّ وَ مَتَاعٌ إِلَىٰ حِينٍ

Transliteration: Fa azallahuma-sh-Shaytanu anha fa akhraja-humā mim-mā kānā fīh, wa qulnā ihbitū bā'dukum liba'din aduwun, wa lakum fil-arżi mustaqarrun wa matā'un ilā ḥīn.

Translation: But Satan caused them to slip out of it and removed them from that [condition] in which they had been. And We said, "Go down, [all of you], as enemies to one another, and you will have upon the earth a place of settlement and provision for a time."

Common Interpretation: This verse narrates the story of Adam and Eve being deceived by Satan and subsequently being expelled from the state of bliss in Paradise. They are directed to descend to the Earth,

The Quranic narrative on Human Evolution

where they will face enmity and challenges but also find a place of settlement and sustenance for a designated period.

Chapter 2, Al-Baqarah, Verse 37:

فَتَلَقَّىٰ اٰدَمُ مِنْ رَّبِّهٖ كَلِمٰتٍ فَتَابَ عَلَيْهِ ۚ إِنَّهٗ هُوَ التَّوَّابُ الرَّحِيْمُ

Transliteration: Fata-laqqa Adamu min rabbihi kalimatin fatāba alayhi innahu huwa at-Tawwāb ur-Raḥīm.

Translation: Then Adam received from his Lord [some] words, and He accepted his repentance. Indeed, it is He who is the Accepting of Repentance, the Merciful.

Common Interpretation: This verse describes the repentance of Adam after he received words from his Lord. It emphasizes the merciful nature of Allah, who is quick to accept repentance and forgive those who turn back to Him.

Chapter 2, Al-Baqarah, Verse 38:

قُلْنَا اهْبِطُوْا مِنْهَا جَمِيْعًا ۚ فَاِمَّا يَاْتِيَنَّكُمْ مِّنِّىْ هُدًى فَمَنْ تَبِعَ هُدَايَ فَلَا خَوْفٌ عَلَيْهِمْ وَلَا هُمْ يَحْزَنُوْنَ

Transliteration: Qulnā ihbitū minhā jamī'an, fa'imma ya'tiyannakum minnī hudan, faman tabi'a hudāya falā khawfun alayhim walā hum yahzanoon.

Translation: We said, Go down from it, all of you. And when guidance comes to you from Me, whoever follows My guidance, there will be no fear concerning them, nor will they grieve.

Common Interpretation: In this verse, Allah commands all of them to descend from Paradise to the Earth. The verse emphasizes that if guidance is received from Allah and followed, there will be no fear or sorrow for those who adhere to His guidance. It indicates the

82

possibility of redemption and a path to overcome the consequences of the earlier disobedience.

Chapter 2, Al-Baqarah, Verse 57:

وَ ظَلَّلْنَا عَلَيْكُمُ الْغَمَامَ وَ اَنْزَلْنَا عَلَيْكُمُ الْمَنَّ وَ السَّلْوٰى ۚ كُلُوْا مِنْ طَيِّبٰتِ مَا رَزَقْنٰكُمْ ۖ وَ مَا ظَلَمُوْنَا وَ لٰكِنْ كَانُوْۤا اَنْفُسَهُمْ يَظْلِمُوْنَ

Transliteration: Wa zallalnā alaikumu al-ghamāma wa anzalnā alaikumu al-manna wa as-salwā. Kulū min tayyibāti mā razaqnākum. Wa mā ẓalamūnā walākin kānū anfusahum yaẓlimūn.

Translation: And We shaded you with clouds and sent down to you manna and quails, [saying], "Eat from the good things with which We have provided you." And they wronged Us not – but they were [only] wronging themselves.

Interpretation: This verse refers to God's blessings upon the Children of Israel during their journey in the wilderness, where they were provided with sustenance in the form of manna and quails. The verse emphasizes that they were not wronging God but rather wronging themselves by their disobedience and ingratitude.

Chapter 2, Al-Baqarah, Verse 60:

وَ اِذِ اسْتَسْقٰى مُوْسٰى لِقَوْمِهٖ فَقُلْنَا اضْرِبْ بِعَصَاكَ الْحَجَرَ ۖ فَانْفَجَرَتْ مِنْهُ اثْنَتَا عَشْرَةَ عَيْنًا ۚ قَدْ عَلِمَ كُلُّ اُنَاسٍ مَّشْرَبَهُمْ ۚ كُلُوْا وَ اشْرَبُوْا مِنْ رِزْقِ اللهِ وَ لَا تَعْثَوْا فِى الْاَرْضِ مُفْسِدِيْنَ

Transliteration: Wa-ithistasqa moosa liqawmihi faqulna idrib bi'aṣāka al-ḥajar fanfajarat minhusnata ashrata aynan qad alima kullu unasin mashrabahum kuloo waishraboo min rizqi Allahi walā ta'thaw fī al-arḍi mufsidīn

Translation: And [recall] when Moses prayed for water for his people, so We said, "Strike with your staff the stone." And there gushed forth from it twelve springs, and every people knew its

The Quranic narrative on Human Evolution

watering place. "Eat and drink from the provision of Allah, and do not commit abuse on the earth, spreading corruption."

Interpretation: This verse refers to the incident when the Prophet Musa (Moses) sought water for his people in the desert. In response to his supplication, God instructed him to strike a rock with his staff, resulting in twelve springs gushing forth, providing water for each of the twelve tribes. The verse encourages gratitude for the sustenance provided by Allah and emphasizes responsible use of resources without causing corruption on the earth. It underscores the importance of utilizing God's provisions while maintaining ethical conduct and avoiding harm to the environment.

Chapter 2, Al-Baqarah, Verse 61:

وَ اِذْ قُلْتُمْ يٰمُوْسٰى لَنْ نَصْبِرَ عَلٰى طَعَامٍ وَّاحِدٍ فَادْعُ لَنَا رَبَّكَ يُخْرِجْ لَنَا مِمَّا تُنْبِتُ الْاَرْضُ مِنْ بَقْلِهَا وَ قِثَّآئِهَا وَ فُوْمِهَا وَ عَدَسِهَا وَ بَصَلِهَا ۖ قَالَ اَتَسْتَبْدِلُوْنَ الَّذِىْ هُوَ اَدْنٰى بِالَّذِىْ هُوَ خَيْرٌ ۖ اِهْبِطُوْا مِصْرًا فَاِنَّ لَكُمْ مَّا سَاَلْتُمْ ۖ وَ ضُرِبَتْ عَلَيْهِمُ الذِّلَّةُ وَ الْمَسْكَنَةُ ۖ وَ بَآءُوْ بِغَضَبٍ مِّنَ اللّٰهِ ۖ ذٰلِكَ بِاَنَّهُمْ كَانُوْا يَكْفُرُوْنَ بِاٰيٰتِ اللّٰهِ وَ يَقْتُلُوْنَ النَّبِيّٖنَ بِغَيْرِ الْحَقِّ ۖ ذٰلِكَ بِمَا عَصَوْا وَّ كَانُوْا يَعْتَدُوْنَ

Transliteration: Wa idh qultum yā Mūsā lan naṣbira alā ṭa'āmin wāḥidin fad'u lanā rabbaka yukhrij lanā mimmā tunbiḍu al-arḍu min baqlihā wa qiththā'ihā wa fuūmihā wa adasihā wa baṣalihā. Qāla atas-tabdilūnal-ladhī huwa adna bil-ladhī huwa khayrun? Ihbiṭū Miṣran fa'inna lakum mā sā'altum. Wa ḍuribat alayhimuẓ-ẓillatu wal-maskanatu wabā'ū bighaḍabin minallāhi. Dhālika bi-annahum kānū yakfurūna bi-āyātillāhi wayaqtulūnan-nabīyīna bighayril-ḥaqqi. Dhālika bimā aṣawwā kānū ya'tadūna.

Translation: And [recall] when you said, "O Moses, we can never endure one [kind of] food. So call upon your Lord to bring forth for us from the earth its green herbs and its cucumbers and its garlic and its lentils and its onions." [Moses] said, "Would you exchange what is better for what is less? Go into [any] settlement and indeed, you will have what you have asked." And they were covered with humiliation and poverty and returned with anger from Allah [upon them]. That was because they [repeatedly] disbelieved in the signs of Allah and killed the prophets without right. That was because they disobeyed and were [habitually] transgressing.

Common Interpretation: This verse serves as a lesson about gratitude, contentment, and the consequences of disobedience to divine guidance. When the Children of Israel expressed their discontent to Moses, stating that they could not endure a single type of food, they requested him to ask Allah to provide them with various crops such as grains, cucumbers, garlic, lentils, and onions. Moses questioned their willingness to exchange what they considered inferior with what was better, but despite his warning, they persisted. As a consequence of their ingratitude and disobedience, they were struck with humiliation, destitution, and the wrath of Allah. This punishment befell them because they habitually rejected the signs of Allah, unjustly killed prophets, and transgressed against His commandments.

Chapter 2, Al-Baqarah, Part of Verse 187:

أُحِلَّ لَكُمْ لَيْلَةَ الصِّيَامِ الرَّفَثُ إِلَى نِسَآئِكُمْ ۚ هُنَّ لِبَاسٌ لَّكُمْ وَأَنتُمْ لِبَاسٌ لَّهُنَّ

Transliteration: Uhilla lakum laylatas Siyaamir rafasu ilaa nisaaa'ikum; hunna libaasullakum wa antum libaasullahunn.

Translation: It has been made permissible for you the night preceding fasting to go to your wives [for sexual relations]. They are a clothing for you and you are a clothing for them.

Common Interpretation: This verse metaphorically depicts the relationship between husband and wife as being akin to clothing, suggesting an intimate and inseparable connection. Just as clothing is close to the body without any barrier, spouses are meant to be intimately close to each other, providing support, comfort, and protection.

Chapter 3, Ale-Imran, Verse 33:

إِنَّ اللهَ اصْطَفَى أَدَمَ وَ نُوْحًا وَّ اٰلَ اِبْرٰهِيْمَ وَ اٰلَ عِمْرٰنَ عَلَى الْعٰلَمِيْنَ

Transliteration: Inna Allaha istafa Adama waNoohan wa-ala Ibraheema wa-ala Imrana alā al-ālamīna

Translation: Indeed, Allah chose Adam and Noah and the family of Abraham and the family of Imran over the worlds.

Common Interpretation: This verse highlights the special selection by Allah of certain individuals and their families, including Adam, Noah, the family of Abraham, and the family of 'Imran, for honorable roles and responsibilities among the people of the world.

Chapter 3, Ale-Imran, Verse 59:

إِنَّ مَثَلَ عِيْسٰى عِنْدَ اللهِ كَمَثَلِ اٰدَمَ ۚ خَلَقَهُ مِنْ تُرَابٍ ثُمَّ قَالَ لَهٗ كُنْ فَيَكُوْنُ

Transliteration: Inna mathala Isa inda Allahi kamathali Adam; khalaqahu min turabin thumma qala lahu kun fayakoon.

Translation: Indeed, the example of Jesus to Allah is like that of Adam. He created Him from dust; then He said to him, "Be," and he was.

Common Interpretation: This verse emphasizes the miraculous creation of both Jesus (Isa) and Adam by Allah. It draws a parallel between them, highlighting the divine power to create through the command "Be," indicating the profound creative ability of God.

Chapter 5, Al-Maida, Verse 27:

وَ اتْلُ عَلَيْهِمْ نَبَاَ ابْنَىْ اٰدَمَ بِالْحَقِّ ۘ اِذْ قَرَّبَا قُرْبَانًا فَتُقُبِّلَ مِنْ اَحَدِهِمَا وَ لَمْ يُتَقَبَّلْ مِنَ الْاٰخَرِ ؕ قَالَ لَاَقْتُلَنَّكَ ؕ قَالَ اِنَّمَا يَتَقَبَّلُ اللهُ مِنَ الْمُتَّقِيْنَ

Transliteration: Wa otlu alayhim naba'abnai Adama bilhaqqi, iz qarraba qurbanan fatoqubbila min ahadihima wa lam yutqabbal min al-akhir. Qala la-aqtulannak. Qala innama yataqabbalu Allahu min al-muttaqin.

Translation: And recite to them the story of Adam's two sons in truth when they both offered a sacrifice [to Allah], and it was accepted from one of them but was not accepted from the other. Said [the latter], "I will surely kill you." Said [the former], "Indeed, Allah only accepts from the righteous [who fear Him]."

Interpretation: This verse recounts the story of Adam's two sons. They both offered sacrifices to Allah, but only one's offering was accepted. This led to jealousy and conflict, with one son expressing an intention to kill the other. The verse highlights the importance of piety, as Allah accepts offerings from the righteous. The narrative serves as a lesson on righteousness, forgiveness, and the consequences of envy.

Chapter 7, Al-Araaf, Verse 16:

قَالَ فَبِمَآ أَغْوَيْتَنِيْ لَأَقْعُدَنَّ لَهُمْ صِرَاطَكَ الْمُسْتَقِيْمَ

Transliteration: Qaala fabimaa aghwaytanee la-aq'udanna lahum siraataka alm-mustaqeem.

Translation: [Iblees] said, "Because You have put me in error, I will surely sit in wait for them on Your straight path."

Common Interpretation: In this verse, Iblees blames Allah for leading him astray and, in response, declares his intent to lie in wait on the straight path of humanity, seeking to misguide them. The verse illustrates Satan's determination to lead people astray and his commitment to obstructing their path to righteousness.

Chapter 7, Al-Araaf, Verse 17:

ثُمَّ لَآتِيَنَّهُمْ مِّنْ بَيْنِ اَيْدِيْهِمْ وَ مِنْ خَلْفِهِمْ وَ عَنْ اَيْمَانِهِمْ وَ عَنْ شَمَآئِلِهِمْ ۖ وَ لَا تَجِدُ اَكْثَرَهُمْ شٰكِرِيْنَ

Transliteration: Thumma laatiyannahum min bayni aydeehim wa min khalfihim wa an aimanihim wa an shama'ilihim, wa laa tajidu aktharahum shakireen.

Translation: Then I will come to them from before them and from behind them and on their right and on their left, and You will not find most of them grateful [to You]."

Common Interpretation: In this verse, Iblees outlines his strategy to approach humans from all sides—before, behind, right, and left—seeking to lead them away from gratitude to Allah. The verse illustrates the comprehensive and persistent nature of Satan's efforts to misguide humanity, emphasizing that many will be ungrateful despite the numerous blessings bestowed upon them.

Chapter 7, Al-Araaf, Verse 19:

وَ يَٰٓـَٔادَمُ اسْكُنْ أَنْتَ وَ زَوْجُكَ الْجَنَّةَ فَكُلَا مِنْ حَيْثُ شِئْتُمَا وَ لَا تَقْرَبَا هَٰذِهِ الشَّجَرَةَ فَتَكُونَا مِنَ الظَّٰلِمِينَ

Transliteration: Wa yā Ādamu uskun anta wazawjuka al-jannata fakulā min haythu shi'tumā walā taqrabā hādhihi ash-shajarata fatakūnā mina az-zālimīn.

Translation: And "O Adam, dwell, you and your wife, in Paradise and eat from wherever you will but do not approach this tree, lest you be among the wrongdoers."

Common Interpretation: This verse recounts that God instructs Adam and his wife to reside in Paradise, permitting them to eat freely from the garden except for the fruit of a specific tree. The warning is given to avoid that particular tree, as disobedience would result in wrongdoing.

Chapter 7, Al-Araaf, Verse 20:

فَوَسْوَسَ لَهُمَا الشَّيْطَٰنُ لِيُبْدِىَ لَهُمَا مَا وُۥرِىَ عَنْهُمَا مِنْ سَوْءَاٰتِهِمَا وَ قَالَ مَا نَهٰىكُمَا رَبُّكُمَا عَنْ هٰذِهِ الشَّجَرَةِ اِلَّآ اَنْ تَكُوْنَا مَلَكَيْنِ اَوْ تَكُوْنَا مِنَ الْخٰلِدِيْنَ

Transliteration: Fa-waswasa lahumā ash-shayṭānu liyubdiya lahumā mā-wūriya anhumā min sawātihimā wa-qāla mā nahākumā rabbukumā an hādhihi ash-shajarati illā an takūnā malakayni aw takūnā mina al-khālidīn.

Translation: But Satan whispered to them to make apparent to them that which was concealed from them of their private parts. He said, "Your Lord did not forbid you this tree except that you become angels or become of the immortal."

Common Interpretation: This verse highlights that Satan whispered to Adam and his wife, suggesting that eating from the forbidden tree would reveal their hidden qualities and that God only prohibited it to prevent them from becoming angels or attaining immortality.

Chapter 7, Al-Araaf, Verse 24:

قَالَ اهْبِطُوْا بَعْضُكُمْ لِبَعْضٍ عَدُوٌّ ۚ وَ لَكُمْ فِى الْاَرْضِ مُسْتَقَرٌّ وَّ مَتَاعٌ اِلٰى حِيْنٍ

Transliteration: Qala ihbitu ba'dukum liba'din aduwun, wa lakum fil-ardi mustaqarrun wa mata'un ila heen.

Translation: [Allah] said, "Descend, being to one another enemies. And for you on the earth is a place of settlement and enjoyment [i.e., provision] for a time."

Common Interpretation: This verse highlights that God commanded Adam and his wife to descend from Paradise, becoming adversaries to each other. They were granted a place on Earth for residence and enjoyment, with guidance provided by God. Those who follow His guidance will not go astray nor suffer.

The Quranic narrative on Human Evolution

Chapter 7, Al-Araaf, Verse 27:

يَٰبَنِىٓ ءَادَمَ لَا يَفْتِنَنَّكُمُ ٱلشَّيْطَٰنُ كَمَآ أَخْرَجَ أَبَوَيْكُم مِّنَ ٱلْجَنَّةِ يَنزِعُ عَنْهُمَا لِبَاسَهُمَا لِيُرِيَهُمَا سَوْءَٰتِهِمَآ إِنَّهُۥ يَرَىٰكُمْ هُوَ وَقَبِيلُهُۥ مِنْ حَيْثُ لَا تَرَوْنَهُمْ إِنَّا جَعَلْنَا ٱلشَّيَٰطِينَ أَوْلِيَآءَ لِلَّذِينَ لَا يُؤْمِنُونَ

Transliteration: Ya Bani Adama, laa yaftinnannakumu ash-shaytanu kama akhraja abawaikum min al-jannati yanzi'u anhumaa libaasahumaa liyuriyahumaa saw'aatihimaa, innahu yaraakum huwa waqabeeluhu min haythu laa tarawnahum. Inna ja'alna ash-shayateena awliyaa'a lilladheena laa yu'minoon.

Translation: O children of Adam! Let not Satan tempt you as he removed your parents from Paradise, stripping them of their clothing to show them their private parts. Indeed, he sees you, he and his tribe, from where you do not see them. Indeed, We have made the devils allies to those who do not believe.

Common Interpretation: This verse warns the children of Adam against the temptations of Satan, recounting the incident of how Satan led their parents (Adam and Eve) astray in Paradise. It underscores the constant presence of Satan and his allies, who observe humans even when not visible to them. The devils are described as allies to those who reject faith.

Chapter 15, Al-Hijr, Verse 39:

قَالَ رَبِّ بِمَآ أَغْوَيْتَنِى لَأُزَيِّنَنَّ لَهُمْ فِى ٱلْأَرْضِ وَ لَأُغْوِيَنَّهُمْ أَجْمَعِينَ

Transliteration: Qaala Rabbi bimaa aghwaytanee la-uzayyi-nanna lahum fil-ardi wa la-ughwiyan-nahum ajma'een.

Translation: [Iblees] said, "My Lord, because You have put me in error, I will surely make [disobedience] attractive to them on earth, and I will mislead them all."

Common Interpretation: In this verse, Iblees acknowledges his deviation and requests permission from Allah to mislead humanity by making disobedience appealing to them on earth, illustrating Satan's determination to exploit humans and tempt them away from righteousness.

Chapter 18, Al-Kahf, Verse 50:

وَ إِذْ قُلْنَا لِلْمَلَٰٓئِكَةِ اسْجُدُوْا لِاٰدَمَ فَسَجَدُوْٓا اِلَّآ اِبْلِيْسَ ۚ كَانَ مِنَ الْجِنِّ فَفَسَقَ عَنْ اَمْرِ رَبِّهٖ ۗ اَفَتَتَّخِذُوْنَهٗ وَ ذُرِّيَّتَهٗٓ اَوْلِيَآءَ مِنْ دُوْنِيْ وَ هُمْ لَكُمْ عَدُوٌّ ۗ بِئْسَ لِلظّٰلِمِيْنَ بَدَلًا

Transliteration: Wa-idh qulnā lil-malā'ikatisjudū li-ādama fasajadū illā iblīsa, kāna minal-jinni fafasaqa an amri rabbihi. Afatattakhidhūnahu wa dhurriyyatahu awliyā'a min dūnī wa hum lakum aduwwun. Bi'sa liẓālimīna badalā.

Translation: And [mention] when We said to the angels, "Prostrate to Adam," and they prostrated, except for Iblis. He was of the jinn and departed from the command of his Lord. Then will you take him and his descendants as allies other than Me while they are enemies to you? Wretched it is for the wrongdoers as an exchange.

Common Interpretation: This verse recounts the command to the angels to prostrate to Adam and the disobedience of Iblis (Satan), who refused to prostrate. It also highlights the enmity between Iblis and humanity and warns against taking Iblis and his descendants as allies.

Chapter 19, Maryam, Verse 58:

أُولٰٓئِكَ الَّذِيْنَ اَنْعَمَ اللّٰهُ عَلَيْهِمْ مِّنَ النَّبِيّٖنَ مِنْ ذُرِّيَّةِ اٰدَمَ ۗ وَ مِمَّنْ حَمَلْنَا مَعَ نُوْحٍ ۗ وَّ مِنْ ذُرِّيَّةِ اِبْرٰهِيْمَ وَ اِسْرَآءِيْلَ ۗ وَ مِمَّنْ هَدَيْنَا وَ اجْتَبَيْنَا ۗ اِذَا تُتْلٰى عَلَيْهِمْ اٰيٰتُ الرَّحْمٰنِ خَرُّوْا سُجَّدًا وَّ بُكِيًّا

Transliteration: Ulaa'ika al-lazeena an'amal laahu alaihim minannabiyyeena min zurriyyati Aadama wa mimman hamalna ma'a Nuuhin wa min zurriyyati Ibraaheema wa Israa'eela wa mimman hadayna wa ajtabayna. Iza tutla alaihim aayaat-ur-rahmaan kharrussujjadaw-wabukiyaa.

Translation: Those are the ones upon whom Allah has bestowed favor from among the prophets of the descendants of Adam and of those We carried [in the ship] with Noah, and of the descendants of Abraham and Israel, and of those whom We guided and chose. When the verses of the Most Merciful were recited to them, they fell in prostration and weeping.

Common Interpretation: This verse acknowledges those individuals favored by Allah, specifically from the lineages of prophets tracing back to Adam. It mentions the descendants of Noah, Abraham, and Israel, as well as those guided and chosen by Allah. When these individuals hear the verses of the Most Merciful, they humbly fall into prostration and express their deep emotions through tears. The verse emphasizes the reverence and humility exhibited by those who have been recipients of divine guidance and favor across various prophetic lineages.

Chapter 20, Taha, Verse 119:

وَ أَنَّكَ لَا تَظْمَؤُا فِيهَا وَ لَا تَضْحٰى

Transliteration: Wa annaka la tazma'u feeha wa la tadha.

Translation: And that you will not be thirsty therein nor will you be hot from the sun.

Common Interpretation: This verse refers to the garden where Adam and Eve were put in for the test.

Chapter 20, Taha, Verse 122:

ثُمَّ اجْتَبٰهُ رَبُّهُ فَتَابَ عَلَيْهِ وَ هَدٰى

Transliteration: Thumma ijtabāhu rabbuhu fatāba alayhi wa hadā.

Translation: Then his Lord chose him, and turned towards him with forgiveness, and granted him guidance.

Common Interpretation: This verse refers to the story of Adam and how his Lord, after he repented, chose him, forgave him, and guided him.

Chapter 20, Taha, Verse 121:

فَأَكَلَا مِنْهَا فَبَدَتْ لَهُمَا سَوْآتُهُمَا وَ طَفِقَا يَخْصِفَانِ عَلَيْهِمَا مِنْ وَرَقِ الْجَنَّةِ ۚ وَ عَصَىٰ أَدَمُ رَبَّهُ فَغَوَىٰ

Transliteration: Fa akalaa minhaa fabadat lahumaa saw aatuhumaa wa tafiqaa yakhsifaani a'laihimaa minw waraqil jannah; wa a'saaa Aadamu Rabbahoo faghawaa

Translation: And they [i.e., Adam and his wife] ate of it, and their private parts became apparent to them, and they began to fasten over themselves from the leaves of Paradise. And Adam disobeyed his Lord and erred.

Common Interpretation: This verse refers to Adam and Eve. It describes the moment when they ate from the forbidden tree after being tempted by Satan. As a consequence, they became naked and tried to cover themselves with leaves from Paradise. This act of disobedience led to their expulsion from Paradise.

Chapter 20, Taha, Verse 123:

قَالَ اهْبِطَا مِنْهَا جَمِيعًا بَعْضُكُمْ لِبَعْضٍ عَدُوٌّ ۖ فَإِمَّا يَأْتِيَنَّكُمْ مِنِّي هُدًى ۙ فَمَنِ اتَّبَعَ هُدَايَ فَلَا يَضِلُّ وَ لَا يَشْقَىٰ

Transliteration: Qāla ihbiṭā minhā jamī'an, ba'ḍukum liba'ḍin 'aduww, fa-immā yatīyannakum mīnnee hudan, famani ittaba'a hudāya falā yaḍillu walā yashqā.

Translation: He (Allah) said, Go down from it, all of you. And when guidance comes to you from Me, whoever follows My guidance, there will be no fear concerning them, nor will they grieve.

The Quranic narrative on Human Evolution

Common Interpretation: In this verse, Allah is recounting the aftermath of the incident in the Garden involving Adam and Eve. After their disobedience, Allah instructs them to descend from the Garden, and He acknowledges that enmity will exist among them on Earth. However, Allah reassures them that if they follow His guidance when it comes to them, they will be free from fear and sorrow. The verse highlights the importance of guidance from Allah as a source of protection and solace for those who follow it.

Chapter 33, Al-Ahzaab, Verse 6:

اَلنَّبِىُّ اَوْلٰى بِالْمُؤْمِنِيْنَ مِنْ اَنْفُسِهِمْ وَ اَزْوَاجُهٗ اُمَّهٰتُهُمْ ۚ وَ اُولُوا الْاَرْحَامِ بَعْضُهُمْ اَوْلٰى بِبَعْضٍ فِىْ كِتٰبِ اللّٰهِ مِنَ الْمُؤْمِنِيْنَ وَ الْمُهٰجِرِيْنَ اِلَّآ اَنْ تَفْعَلُوْۤا اِلٰٓى اَوْلِيٰٓئِكُمْ مَّعْرُوْفًا ؕ كَانَ ذٰلِكَ فِى الْكِتٰبِ مَسْطُوْرًا

Transliteration: Al-Nabiyyu awla bil-mu'minina min anfusihim wa azwajuhu ummahatuhum. Wa u'lul-arhami ba'duhum awla bil-ba'din fi kitabillahi min al-mu'minina wal-muhajirina illa an taf'alu i'la awliya'i-kum ma'rufan. Kana dhalika fil-kitabi mastura."

Translation: The Prophet is more worthy of the believers than themselves, and his wives are their mothers. And those of [blood] relationships are more entitled [to inheritance] in the decree of Allah than the [other] believers and the emigrants, except that you may do to your close associates a kindness [through bequest]. That was in the Book inscribed.

Common Interpretation: This verse establishes the elevated status of the Prophet Muhammad among the believers, likening him to a protector and guide. It emphasizes the special position of his wives as the mothers of the believers. Additionally, it addresses matters related to inheritance, indicating that relatives have a greater entitlement in the distribution of inheritance, as specified in the Book of Allah.

Chapter Five: Origin

Before delving into the insights offered by the Quran regarding the origins of life, it is essential to consider the scientific perspective on origin of life as cited in Chapter Two of the book, "Science of Human Evolution."

The journey of species' evolution forms the latter part of the narrative, but the roots from which it all began bear significant importance. Strikingly, there are notable similarities between the Quranic account and certain scientific concepts.

It holds great significance that the Quran has extensively explored the origins of life across numerous verses. Despite these revelations spanning over fourteen centuries, contemporary scientific breakthroughs have shed light on invaluable information. This enrichment enhances our comprehension of the Quranic verses, injecting newfound depth into the ongoing discourse. Refer to Group 1, Origin of Life, of Chapter Four: "Extraction of verses."

Nafs-in Wahid (A Single Soul)

The Quran elucidates on the creation of human beings from a "Nafsin Wahida" in Arabic (نَفْسٍ وَاحِدَةٍ). The term "Nafs" in Arabic, often translated as "Soul" or "living organism," and "Wahida" meaning "single," collectively conveys the literal meaning of a single soul or a single living organism. Numerous Quranic verses allude to the origin of human creation from a singular living entity. For instance,

Chapter 4 Verse 1 encompasses, "O mankind, fear your Lord, who created you from one soul."

Chapter 6 Verse 98 encompasses, "And it is He who produced you from one soul."

Chapter 7 Verse 189 includes, "It is He who created you from one soul and created from it its couple."

Chapter 39 Verse 6 includes, "He created you from one soul; then He made from it its couple."

According to the interpretation provided by many Muslim scholars who have translated and explained the Quran, the initial two verses are generally regarded as a narrative of the creation of the human being from Adam. The subsequent two verses are commonly understood to depict the creation of Eve, the counterpart, from Adam.

Nevertheless, it's crucial to observe that these verses, which delve into the origins of humankind, do not explicitly mention the names of Adam or Eve. The interpretation offered by Muslim scholars, connecting these verses to the creation of Adam and Eve, might originate due to various reasons.

The prevailing narrative of Adam and Eve portrays them as the first human beings, thus predetermining the outcomes of humankind's creation starting with them. Another reason might be that delving into the philosophical depths of sacred scriptures and deriving differing interpretations from established narratives often encounters resistance in certain societies. Questions may arise as to why such interpretations have not surfaced over the course of many centuries and why the concept has not been understood in this manner before. Furthermore, a deficiency in scientific knowledge or the inclination to arrive at swift conclusions may also contribute to these factors.

Within the Quran, the term "Nafs" is recurrent, its significance echoing across several verses. The meanings encapsulated by "Nafs" are both diverse and nuanced, with interpretation hinging upon the specific context in which the term is employed. Predominantly, this word is utilized to denote human beings, delving into the intricacies of human nature. Additionally, "Nafs" serves as a signifier for desire, ego, and superego or conscience, particularly within the realm of human behavior.

Remarkably, the word extends its semantic reach to encompass the divine self, representing God. This transcendent usage is notably exemplified in Chapter 3 Verses 28 and 30, as well as Chapter 6

Verses 12 and 54. In these instances, "Nafs" takes on a profound and sacred dimension, signifying God's self.

A verse holds a prominent place in Islamic understanding, universally acknowledged by Muslims as a reminder of the inherent mortality that applies to all living creatures. The opening segment of Chapter 29 Verse 57 reads as follows:

Arabic: كُلُّ نَفْسٍ ذَآئِقَةُ الْمَوْتِ

Transliteration: Kullu nafsin thā'iqatul-mawt

Translation: Every soul will taste death.

There exists a consensus among Muslim scholars, coupled with widespread recognition, that the term "Nafs" in this verse encompasses all living beings. Consequently, this verse extends an invitation to delve into the broader implications of the term "Nafs" in its literal sense, facilitating a more comprehensive understanding of the verses, particularly in relation to the creation of human beings.

Should one abstain from interpreting the term "Nafs" as specifically alluding to Adam in this particular verse, it becomes apparent that there is no inherently compelling reason to extend this interpretation to all the verses containing the term "Nafsin Wahida." Unless one considers the factors previously mentioned that prompted Muslim scholars to interpret "Nafs" as referring to Adam rather than a living being or living organism.

Remarkably, Dr. Tahir-ul-Qadri, a highly esteemed Muslim scholar, has translated the term "Nafsin Wahida" as "single cell" (Chapter 6 Verse 98) and "single living cell" (Chapter 39 Verse 6). This translation adds a compelling dimension to the interpretation, emphasizing that these verses may not be referring to the origin of the human being from Adam but rather from a single cell.

Consequently, the question naturally arises: How does the term "Nafsin Wahid" correlate with scientific understanding?

The Quranic narrative on Human Evolution

In juxtaposition to established scientific discoveries and theories, the Panspermia hypothesis, though reminiscent of a theological perspective, suggests an alternative narrative: life did not originate on Earth but was rather introduced from an external source, albeit with differing external origins in various interpretations. However, when examined through the lens of the Abiogenesis hypothesis, the Quranic concept of a singular Nafs becomes more readily understandable.

The Quran is thought to provide insights into the genesis of the human being from a single cellular entity, which undergoes subsequent division, ultimately giving rise to its counterpart. In scientific terms, this phenomenon aligns with asexual reproduction, specifically known as mitosis.

Adding to the intrigue is the conspicuous absence of any reference to the gender of Nafs or its pair within these verses. The Arabic term "Zaoj" (زَوْجَهَا) is employed, whose English equivalent is "spouse." This corresponds to scientific understanding, which suggests that the differentiation of sexes or the classification of gender took place much later, perhaps billions of years after the initial events being described.

In the realm of scientific inquiry, there exists a prevailing humility when attempting to understand the fundamental "why" behind observed phenomena. Scientific hypotheses excel at elucidating the "what" and "how" of events but often remain silent on the "why." In this context, the Quran boldly steps into this intellectual void, explicitly affirming the reasons behind these occurrences. It asserts that these events unfolded as an outcome of divine will, guided by God's intention to shape humanity from a single cell and, subsequently, bring forth its counterpart.

Furthermore, Abiogenesis delves deeply into the intricacies of life's origin, exploring the details of how it occurred and what ingredients founded this initial form of life. Remarkably, the Quran also contributes to this profound discourse, offering its insights on the subject.

Creation from Clay

Certain scientists propose Abiogenesis as a phenomenon that might have occurred in the vicinity of hydrothermal vents. These geological features expel high-temperature, mineral-rich fluids into the surrounding seawater. Notably, there is a line of thought among these scientists suggesting that specific minerals, particularly clay minerals, could have played a role in facilitating the formation and concentration of organic molecules.

Refer to Group 2, Creation with Clay, of Chapter Four: "Extraction of verses." The Quran states in Chapter 32 Verse 7, *"[God] who perfected everything which He created and began the creation of man from clay."* In this verse, the Arabic word "Badaa" (بَدَأَ) is employed, which translates to "began" or "initiated" in English. Also, in the Quran, Chapter 71 Verse 17, a clear proclamation is made that human beings are fashioned as a product of the earth's growth, *"And Allah has caused you to grow from the earth a [progressive] growth."*

The Quran asserts two fundamental concepts: first, the creation of humankind from a single Nafs, which can also be described as a single cell, and second, the initiation of the creation of humankind with clay. It is quite evident that these two notions are intricately linked, suggesting that the formation of the single Nafs indeed began with clay.

The Quran has described the formation of single cell from clay in various stages, including dust, clay, mud, extract of clay, sticky clay, and hard clay. These stages represent the systematic process by which the creation of a single Nafs was initiated that ultimately led to the formation of human beings.

Following are the different terms portraying each stage. However, it is important to note that these verses may not align with a specific chronological sequence, as the Quran does not explicitly delineate the order of these stages.

The Quranic narrative on Human Evolution

Dust (Turaab): Chapter 18 Verse 37 includes *"We created you from dust (Turaab)."*

Clay (Teen): Chapter 06 Verse 02 includes, *"He [God] is the one who created you from clay (Teen)."*

Mud (Salsaal-in min Hama-in Masnoon): Chapter 15 Verse 26 states, *"And We certainly created man out of clay from an altered black mud (Salsaal-in min Hama-in Masnoon)."*

Extract of Clay (Sulaal): Chapter 23 Verse 12 states, *"And We certainly created man from an extract of clay (Sulaal)."*

Sticky Clay (Teen-in Laazib): Chapter 37 Verse 11 includes, *"Indeed, We created them [i.e., disbelievers] from sticky clay (Teen-in Laazib)."*

Hard Clay (Salsaal-in Kalfakhaar): Chapter 55 Verse 14 states, *"He has created man from dry, rotten clay like the potter's (Salsaal-in Kalfakhaar)."*

These verses elucidate the progressive stages in the formation of the Nafs-in Wahid or the single-cell organism. It's important to note that while many Muslim scholars interpret these verses in the context of Adam's creation but the Quran's text itself does not exclusively associate this process with Adam. Nonetheless, given that Adam is a human being, these verses are equally applicable to him as they are to any other human being.

Creation with Water

Just like the theory of Abiogenesis postulates, another essential element in the creation process is water. All the verses noted in Group 3, Creation with Water, of Chapter Four: "Extraction of verses," elucidate the role of water in the act of creation.

Notably, Chapter 24 Verse 45 primarily addresses the creation of moving creatures (دَآبَّة) with water, *"And God created every moving*

100

creature from water. So of them is that which moves upon its belly, and of them is that which moves upon two legs, and of them is that which moves upon four. God creates what He wills. Indeed, God is over all things competent."

Similarly, Chapter 25 Verse 54 discusses the creation of humans (بَشَرً) with water, *"And it is He who has created from water a human being and made him [a relative by] lineage and marriage. And ever is your Lord competent [concerning creation]."* And, Chapter 21 Verse 30 addresses the creation of all living beings (كُلَّ شَىْءٍ) with water. The verse encompasses, *"(God) made from water every living thing?"*

The Quran firmly asserts that all living entities on Earth share a fundamental and essential constituent of life, which is water. It is important to emphasize that when the Quran expounds upon the intricacies of forming a single cell and fashioning life from clay, it exclusively relates to the creation of human beings. In contrast, water is distinctly highlighted as the life-giving source for both humans and other creatures, underlining its universal significance as the sustainer of life for all living beings.

In Chapter 21 Verse 30 of the Quran, it is stated, *"Do those who disbelieve not see that the heavens and the earth were a [closed] piece [ratqan], and We opened them and made from water every living thing? Then will they not believe?"* While this verse does contain an affirmative notion that could be linked to the Last Universal Common Ancestor (LUCA) concept due to the reference to water in the context of the creation of all living beings, it would be somewhat speculative to assert a direct connection. The primary focus of the verse is on the creation of the Earth and the vital role of water in the emergence of life on our planet. Therefore, it may be overly ambitious to employ this verse as definitive evidence in support of the theory of LUCA.

However, it's essential to highlight that there is no explicit Quranic text that opposes the idea of one common ancestor for all living creatures, including humans. Therefore, this concept cannot be classified as "haraam", meaning forbidden or prohibited in Islam. The theory of the Last Universal Common Ancestor (LUCA) should be evaluated on its own merits, without the burden of religious prohibition.

Given that LUCA is postulated as a single-cell organism, it could also be likened to Nafs-in Wahid or a single cell as described in the Quran. However, it is crucial to acknowledge that by accepting this link, one would essentially assert that LUCA served as the inception point of human creation. This interpretation is open to individual belief and should be examined within the context of one's faith and scientific understanding.

In the realm of scientific terminology, "Mitosis" refers to a type of asexual reproduction wherein a cell divides into two identical cells. On the other hand, "Meiosis" is the term employed to describe the process of sexual reproduction. What's intriguing is that the scientific community has yet to provide a definitive answer to the question of how asexual reproduction transitioned into sexual reproduction. Although several theories and hypotheses exist, all pointing toward the attributes of sexual reproduction, they posit that the development of distinct male and female sexes, along with sexual reproduction, could have evolved to enhance genetic diversity and increase the ability to adapt to changing environmental conditions.

Critics proffer more persuasive arguments opposing the transition from asexual to sexual reproduction. Primarily, these critics are aligned with creationist beliefs, subscribing to the notion of divine creation as opposed to the theory of evolution. Here are some of their arguments,

Asexual species continue to exist and thrive today, indicating that sexual reproduction is not universally advantageous from an evolutionary perspective.

Sexual reproduction is costly compared to asexual reproduction. In asexual reproduction, all offspring inherit 100% of their parent's genes. In sexual reproduction, only 50% of an individual's genes are passed on to their offspring. This reduction in genetic relatedness might seem counterproductive from an evolutionary perspective.

The theory of the two-fold cost of sex posits that in sexual reproduction, merely half of the population, specifically the females, partake in the process of reproduction. In asexual populations, conversely, each individual retains the capacity for reproduction. This implication implies that sexual reproduction may exhibit lesser efficiency in the context of population growth.

The sexual reproductive process must be fully operational to effectively convey genetic information and any concurrent genetic modifications to the offspring. Therefore, the transition from asexual to sexual reproduction cannot reasonably be attributed to a gradual, incremental process.

As science currently lacks a comprehensive explanation for the intricacies of the transition from asexual to sexual reproduction, theologians have seized upon this gap as a formidable basis to challenge the very premise of evolutionary theory. Intriguingly, the Quran offers insights into this matter.

Refer to the Group 4, Stages of Creation, of Chapter Four: "Extraction of Verses", of the book. These verses primarily elucidate the various stages of creation. In particular, one verse grabbed an extended period of contemplation to extract its deeper meaning. The first part of Chapter 35 Verse 11 states, *"And God created you from dust, then from a Nutfa (sperm or reproductive cell); then He (God) made you mates."*

The key to understand the true meaning of this verse is to analyze the chronological order of the stages.

1. Creation from dust

2. Then, creation from Nutfa - sexual reproduction

3. Then, pairing up as mates.

From a theological perspective, the widely accepted belief aligns with a significant sequence, distinct from the one mentioned in the Quran, that is,

1. Creation from dust – (that is Adam's creation from dust).

2. Then Pairing up as mates – (by the creation of Eve as Adam's companion).

3. Then creation from Nutfa - sexual reproduction (initiated by Adam and Eve).

Regrettably, numerous scholars engaged in the translation and interpretation of the Quran have tended to overlook the significance of the chronological order of the stages of human creation. Instead, the majority have placed a stronger emphasis on the latter portion of the verse, which expounds upon God's authority over matters of life and death.

The sequence of the creation stages presented in the verse is distinct because it does not link to the creation of Adam. Instead, it delineates the **creation from dust** - the process termed Abiogenesis in scientific terminology, which corresponds to the creation of a single cell. This is succeeded by the act of asexual reproduction known as Mitosis in scientific discourse, representing the process of a single cell reproducing into two identical cells. Then, **the creation from Nutfa** or productive cells, referred to as Meiosis in scientific terms, elucidates the concept of sexual reproduction. In the language of science, this verse clarifies the progression from abiogenesis to asexual reproduction and, ultimately, to sexual reproduction.

The mention of **pairing up of mates** or spouses in the latter part of the verse signifies that, as an outcome of these processes, there emerges a multitude of males and females paired up and engaged in reproduction.

Although science has not yet provided a conclusive theory or hypothesis explaining the shift from mitosis to meiosis, an interesting correlation can be observed between the Quran and scientific principles. The Quran underscores the importance of sperm or reproductive cells in the evolutionary sequence, preceding the complex process of sexual reproduction through mating. However, it doesn't delve into the mechanics of this transition, attributing it to the divine command. It is conceivable that as science advances, it may provide insights that align more harmoniously with the verses of the Quran. Until such a time, it is reasonable to suggest that there is a degree of agreement between the realms of science and the Quran on this matter.

Chapter Six: Bashar

The Quran contains various characteristics and descriptions of human beings, one of which is "Bashar" (بَشَرًا). Typically, it is translated as "human", just like the word "Insaan" (الإنسان). But to grasp the true essence, one must delve into the profound reasoning behind the Quran's careful choice of words.

Bashar is an attribute used in the Quran that pertains to the earthly, animal aspects of human beings, including their instincts and natural inclinations. In this sense, human beings, or Insaan, can be categorized as Bashar due to the presence of these animalistic qualities. However, it's important to note that while human beings can be classified as Bashar, the reverse is not true; Bashar cannot be classified as human beings or Insaan. This distinction underscores the multifaceted nature of human beings and their dual existence as both physical creatures with animalistic traits and spiritual beings with higher moral and intellectual capacities.

Refer to Group 5, Bashar, of Chapter Four: "Extraction of Verses." God informed angels about the creation of a new being. Chapter 15 Verse 28 includes, *"When your lord said to the angels, I [God] am creating Bashar from clay from moulded mud"* and Chapter 38 Verse 71 states, *"When your lord said to the angels, I [God] am about to create Bashar from clay."* These verses highlight the specific creation of Bashar from clay, signifying a distinct phase in the divine creation. It is crucial to distinguish that this is not the stage of creating Insaan but a Bashar.

This was the period when the creation of a single cell, or Nafsin Wahid, was set to commence from the Earth's crust. Angels were informed at this early stage that the creation of a Bashar is about to originate. And it commenced with the creation of this single cell from clay.

In contrast to conventional evolutionary theories, the Quran underscores divine selection over natural selection, emphasizing a deliberate and systematic transformation rather than random mutation.

The Quranic narrative on Human Evolution

However, the Quran aligns with the idea that this transformation is a gradual process, although it doesn't specify the duration of this period.

The creation of a Bashar in the Quranic narrative does not appear to be a step that is directly followed by meiosis or sexual reproduction. The Quran does not describe any intermediate stages between meiosis and the creation of "Bashar." Rather, it alludes to a period in the history of humanity that is considered insignificant. In Chapter 76 Verse 1 of the Quran, it is stated, *"Has there [not] come upon human a period of time when he was not a thing worth of mentioning."*

This verse underscores the idea that there was a time when human beings were in a state of insignificance, and the Quran does not delve into the specific details of this phase. It serves as a reminder of the humble beginnings of human existence, highlighting the transformative power of divine creation. This has been interpreted by Muslim scholars in two primary ways:

Firstly, a common interpretation among scholars is that this period implies the early stages of human development when humans were in the form of a sperm. This interpretation aligns with various verses in the Quran that describe the creation of humans from a drop of sperm.

Secondly, some scholars perceive this period as the moment when all human souls were gathered and made to testify to their acknowledgment of God's lordship. This interpretation is rooted in Quran Chapter 7 Verse 172, where it is mentioned how God gathered the souls of all human beings that were to come to Earth until the end of time. These souls were then made to testify, ensuring that no one can claim ignorance on the Day of Resurrection. The verse states, *And [mention] when your Lord took from the children of Adam - from their loins - their descendants and made them testify of themselves, [saying to them], "Am I not your Lord?" They said, "Yes, we have testified." [This] - lest you should say on the Day of Resurrection, "Indeed, we were of this unaware."*

However, it is important to note that neither of these interpretations fully aligns with the straightforward, literal meaning of the Quranic verse. Precisely because this phase was deemed unworthy of elaboration, it remains conspicuously absent from the Quranic text. In contrast, the two common interpretations offered by scholars are indeed mentioned and described in various verses of the Quran.

Exploring the era identified as "not worth mentioning" in conjunction with scientific findings provides a novel perspective on the Quranic verse. The Quran seems to omit the intermediate stages of evolution between meiosis or sexual reproduction and the creation of Bashar, equated with the scientific classification of Homo sapiens, excluding modern humans categorized as Homo sapiens. This leap seems to bypass all the intermediary stages, encompassing the rest of Primates, Hominids, Hominins or Homos and labelling these stages as not deserving explicit mention within the Quran.

In Chapter 38 Verse 72 of the Quran, God imparts knowledge to the angels regarding the creation of the first human being, stating, *"So when I have proportioned him (Bashar) and breathed into him My [created] Rooh (spirit), then fall down to him in prostration."* The term "proportioned" carries profound significance. Although the Quran does not provide details on the process or specify a time period, it unequivocally indicates that Bashar has undergone stages of proportioning — in other words, a process of evolution — to reach a perfected state, ready to be transformed into a human being.

As mentioned earlier, the Quran deliberately omits explicit mention of the stages of this evolution, deeming it unworthy of detailed discourse. The reason for this choice is better known by divine wisdom.

Human Descent from Apes?

The question of whether our ancestors were apes remains without a definitive answer in the light of the Quran. However, theologians and believers should not face any hindrance in embracing this concept if they consider it the method chosen by God to create human beings, encompassing the stages of evolution, even if it entails a period where

The Quranic narrative on Human Evolution

our ancestors displayed traits similar to apes. The Quran's silence on the matter permits such interpretation, emphasizing the compatibility between faith and scientific understanding.

The hesitation to embrace the idea of apes as potential ancestors of humankind shares a parallel with the reason Satan refused to bow down to Adam, as elucidated in the subsequent chapter of this discourse. Satan, in his arrogance, regarded humans as they progressed through various stages of development after being created with the earth crust, perceiving them as akin to animals of lesser stature than himself. This perspective conflicted with his pride and conscience, preventing him from humbling himself before Adam.

In a similar vein, humans, who perceive themselves as superior to or more elevated than apes, may struggle with the notion of being descendants of apes. This comparison underscores the innate inclination to distinguish ourselves from creatures we consider of lower order, mirroring Satan's dilemma when he regarded humans from a vantage point above.

It is important to recognize that even if the Quran or the Bible would have explicitly declared that humans are descendants of apes, the inherent complexity of human psychology suggests that disbelievers might have seized upon this revelation to criticize and ridicule believers, much as they do today when believers do not adhere to the theory of human descent from apes. The dynamics of belief and disbelief can be influenced by a variety of factors, and such a revelation would likely have been met with resistance and scepticism by some.

Before Charles Darwin proposed his theory of evolution, Muslim scientists and scholars had already explored similar ideas. Interestingly, the concept that humans evolved from apes or monkeys was once known as the "Mohammadan Theory of Evolution," a notion that was opposed by scholars from other faiths. The input of Muslim scientists in this area are discussed in Chapter Two, under the heading "Contribution of Muslim Scientists."

It is noteworthy that there is no historical evidence of significant conflict or controversy arising from the research of Muslim scholars on evolution within the Muslim world. Therefore, it can be inferred that the controversy among Muslims regarding evolution is relatively newer compared to the debates among followers of Christianity or Judaism.

In contemporary times, many Muslim clergy members express opposition to the concept of evolution and consider it Haraam (forbidden) within Islam. However, according to Islamic principles, a notion can only be classified as Haraam if it is explicitly forbidden in the Quran or Hadiths (the sayings of Prophet Muhammad). There exists no evidence that the notion of humans evolving from apes or monkeys was condemned or prohibited within Islam during its first 1200 years, before Charles Darwin suggested the idea that all life shares a common ancestry. Consequently, there appears to be comparatively more leeway within Islam to engage with scientific perspectives on this matter.

Chapter Seven: First Human Being

The subsequent stage in the Quran's account of creation, following the creation of Bashar, is the formation of the first fully developed human being. Refer to Group 6, First Human Being, of Chapter Four: "Extraction of Verses."

Chapter 38 Verses 71 and 72 of the Quran stated, *"[So mention] when your Lord said to the angels, "Indeed, I am going to create a Bashar (animalistic human) from clay. So when I have proportioned it (Bashar) and breathed into it (Bashar) my [created] Rooh (spirit), then fall down to him in prostration."*

It is worth noting that in this particular verse, the absence of Adam's name is intentional, as the verse refers to the time predating the emergence of Bashar, likely during the phase when clay served as the catalyst, marking the inception of Nafsin Wahid or the singular biological organism. It is only later, when the event unfolds, and angels are once again instructed to bow, that the name Adam is introduced, as illustrated in Chapter Verse 34 of the Quran, *"And [mention] when We said to the angels, 'Prostrate to Adam,' and they prostrated."*

The act of proportioning Bashar can be interpreted as a reference to the concept of evolution in scientific terms. However, it is important to note that the Quran does not provide specific details about the stages of this evolutionary process. It designates this period as insignificant in the timeline of human creation, as previously discussed in the book. Therefore, while the Quran may allude to the concept of evolution in a broader sense, it does not offer detailed information about the specific stages involved.

Since the Quran also does not specify the timeframes for the stages of creation in detail, it is challenging to confirm the exact duration between the inception of Bashar, the period of proportioning it, and its transformation to the first human being. The Quran provides a general account of creation but does not offer specific timelines for these events.

The Quranic narrative on Human Evolution

To comprehend the concept of Rooh (spirit) as described in the Quran, it is essential to have an understanding of the different types of divine creation. The Quran indicates two types of divine creation, one of which is Alam-e-Khalq - "خَلْق" (the world of creation), which refers to the act of bringing something into existence, the process of forming and making something. This type of creation is characterized by the passage of time, involving both the origination and evolution of the created being. The various stages of human creation discussed in the Quran, from the single cell to Bashar, exemplify the concept of Khalq, which encompasses the gradual development and evolution of beings.

In contrast, the other type of divine creation is characterized by its immediacy and does not require time for completion. This type of creation occurs as soon as God commands it to be. Chapter 36 Verse 82 of the Quran exemplifies this concept, stating, *"His (God) command is only when He intends a thing that He says to it, "Be", and it is."* This form of divine creation is referred to as Alam-e-Amr – "أَمْر" (World of Command). The Rooh (spirit) is one of the examples in the Quran of creation by God's command, falling into this category of immediate, divine creation, without the need for a gradual process of evolution.

Human beings are a unique example in which both types of divine creation, Alam-e-Khalq (the World of Creation) and Alam-e-Amr (the World of Command), are involved. Bashar, the physical form, is created through Alam-e-Khalq, which include evolution and gradual development. Subsequently, the Rooh (spirit), a creation of Alam-e-Amr, is brought into existence by divine command. The Rooh is then placed within the Bashar, resulting in the creation of a complete human being. This dual nature of creation, with both the physical and spiritual aspects, is a distinctive feature of human existence according to the Quranic perspective.

In this context, the Quran mentions in the first part of Chapter 38 Verse 75, *"The Lord said: O Iblis (Satan), what prevented you from prostrating yourself before him whom I created of My Two Hands."* The metaphor "two hands" is employed to signify the creation of the

The Quranic narrative on Human Evolution

human being through both types of divine creation, Alam-e-Khalq (the World of Creation) and Alam-e-Amr (the World of Command). This underscores the comprehensive nature of human creation, which encompasses both the physical and spiritual aspects.

In the milieu of creation, science primarily explores the material and observable aspects of existence that is Alam-e-Khalq. The Quran often urges its readers to reflect and seek understanding, encouraging the pursuit of knowledge in this realm. However, when it comes to the concept of Rooh, which delves into the spiritual and metaphysical dimensions of creation, the Quran conveys that only a limited comprehension is provided to human beings. This limitation in understanding may be attributed to the fact that the concept of Rooh is closely related to Alam-e-Amr, the realm of divine command and guidance.

Quran mentions an account when people questioned Muhammad (peace be upon Him) about the nature of Rooh. In the Quran Chapter 17 Verse 85 states, *"And they ask you, [O Muhammad], about the Rooh (spirit). Say, "The Rooh is of the affair [i.e., concern] of my Lord. And you [i.e., mankind] have not been given of knowledge except a little."*

While the Quran encourages reflection and understanding in many areas, it acknowledges the boundaries of human knowledge, particularly when it comes to matters beyond the physical world. The concept of Rooh serves as a reminder of the profound spiritual aspects of existence, and it suggests that a complete understanding of this realm remains within the purview of divine knowledge. This distinction highlights the Quran's recognition of both the tangible and intangible aspects of creation, underscoring the notion that certain aspects of existence are beyond the scope of empirical science and are best understood through a lens of faith and spirituality.

Expanding on this concept, we enter a realm that sparks the imagination. In Chapter 79 Verses 1 and 2 of the Quran, a moment is described—the time of death, when angels are tasked with the collection of the Rooh, from a human body, *"By those (angels) that pluck out the soul from depths, and gently take it away."* This imagery portrays the angels delving into the body to extract the spirit. But where do they pluck it from? Is it a fragment of a cell? A component of DNA, perhaps?

Yet, a fundamental distinction emerges—both the cell and DNA are physical and structural elements of Bashar, the pre-human form, while Rooh represents the spiritual essence of a human being. This prompts a profound question, one that transcends the boundaries of empirical knowledge. The nature of Rooh, its origin, and its connection to the physical self remain a mystery, and our understanding is indeed limited in this matter.

Nevertheless, we can discern the profound impact of the Rooh when we compare a Bashar to a fully developed human being. In scientific terminology, the human being in question corresponds to the modern humans also classified as Homo sapiens sapiens, while the Bashar represents a pre-human entity that was on the verge of becoming the first human, classified as Homo sapiens. This comparison allows us to observe the stark contrast between a Bashar without a Rooh and a Bashar imbued with a Rooh.

This transformation is not merely a physical one; it is a profound metaphysical shift that encompasses several key aspects,

Spiritual Essence: The Rooh represents the spiritual essence of a human being. It is considered a divine gift that elevates humans above the material world. The introduction of the Rooh infuses the being with a spiritual dimension, signifying a connection to the divine and a higher purpose in life.

Consciousness: The Rooh brings with it, consciousness. While animals and other living creatures may exhibit basic forms of awareness, human consciousness is distinct. It includes self-awareness, moral consciousness, and the capacity for deep introspection. Humans can ponder questions about the meaning of life, ethics, and the nature of the universe. This consciousness is a product of the Rooh and sets humans apart from other forms of life.

Emotions: With the Rooh comes a rich tapestry of emotions. Human beings experience a wide range of feelings, from joy and love to sorrow and empathy. Emotions are integral to the human experience and are linked to the presence of the Rooh. They play a significant role in shaping human relationships, art, culture, and morality.

Intellect: The Rooh is closely associated with human intellect. It grants humans the ability to reason, think abstractly, and engage in complex problem-solving. This intellectual capacity enables humans to advance in fields like science, philosophy, and art. It is the Rooh that provides the intellectual foundation for human achievements and innovations.

Metaphysical Connection: The Rooh connects human beings to the metaphysical realm. It allows individuals to contemplate questions about the existence of God, the purpose of life, and the nature of the spirit itself. This metaphysical dimension is an essential part of the human experience and is closely tied to the Rooh.

The introduction of the Rooh marks a profound transformation from a Bashar to a human being. The Rooh is the core of human identity, shaping our unique attributes and defining our humanity. It is the bridge between the physical and spiritual dimensions of human existence, reflecting the complex and multifaceted nature of our being.

It is the Rooh that not only transformed a Bashar into a human being but also elevated them to the status of God's deputy on Earth. Chapter 2 Verse 30 of the Quran states, *"And [mention, O Muhammad], when your Lord said to the angels, "Indeed, I will make upon the earth a successive authority." They said, "Will You place upon it one who causes corruption therein and sheds blood, while we exalt You with praise and declare Your perfection?" He [God] said, "Indeed, I know that which you do not know."*

The first segment of the verse signifies that God chose to appoint His Khalifa (خَلِيفَة) or Caliph. The term "Khalifa" is commonly translated as "vicegerent," "steward," or "representative". Designating humans

The Quranic narrative on Human Evolution

as Khalifa on Earth implies entrusting them with a unique role and responsibility.

The concept of being Khalifa of God on Earth implies that,

1- Humans possess free will and the ability to make choices. They are accountable for their actions and are expected to make choices that align with God's guidance.

2- Humans are expected to uphold moral and ethical principles. This includes promoting justice, compassion, and fairness in all aspects of life. They are responsible for maintaining a just and ethical society.

3- Humans are expected to engage in spiritual growth and a relationship with the divine. This includes acts of worship, prayer, and self-purification.

Overall, making humans Khalifa on Earth reflects the idea that humans have a unique status and purpose in the world. They are meant to fulfil a special role as stewards, moral beings, and agents of progress and civilization while being accountable for their actions. This concept emphasizes both the earthly and spiritual dimensions of human existence and their relationship with the Creator.

The second segment of the verse (Chapter 2 Verse 30) delves into the apprehension expressed by the angels regarding the appointment of human beings as Khalifa. In response to God's decision, they posed a question, *"Will you place upon it (Earth) one who causes corruption therein and sheds blood, while we declare Your praise and sanctify You?"*

Muslim scholars have traditionally interpreted the angels' response as a prediction that granting a being the responsibility of becoming God's deputy on Earth may lead to power, privileges, and authority. They expressed concerns that with power often comes the potential for corruption. As for their apprehension about bloodshed, scholars have explained it by referring to the Djinns (also spelled as Jinns or Genies) who were inhabitants of the Earth before humans. They were also given free will and were believed to have caused bloodshed on Earth.

The Quranic narrative on Human Evolution

However, it is essential to note that there are aspects of this interpretation that may not align perfectly with the Quranic text.

It is vital to recognize that the angels' response is not a prediction of future events, but a genuine concern rooted in the ongoing reality. The literal translation of their concern in the verse is, "who corrupts in it and sheds the blood," which indicates that these negative actions were already taking place at the time they expressed their apprehension. Moreover, the comparison between humans and djinns concerning the shedding of blood is not valid, as djinns were never appointed as God's Khalifa on Earth. The angels' concern should be understood as a response to existing circumstances rather than a prophecy of future events.

The angels, in their celestial realm, observed the Bashar or pre-humans on Earth engaged in acts of bloodshed and violence. This observation aligns with the findings in contemporary science, particularly in archaeological and fossil records, which provide substantial evidence of such behaviors among early hominins. These traces of violence and brutality include indicators of interpersonal conflicts, territorial disputes, and even instances of cannibalism.

Through the meticulous examination of ancient skeletal remains, archaeologists and paleoanthropologists have uncovered telltale signs of violence, such as evidence of cranial injuries, fractures, and skeletal trauma. These injuries, often marked by healed or unhealed wounds, offer a glimpse into the harsh and perilous existence of these early human ancestors. In some instances, the presence of weapon-like objects and tools further substantiates the hypothesis of intergroup conflicts and potential violence.

Moreover, findings of cannibalism provide valuable insights into the complex and challenging dynamics of early human societies. Such evidence underscores the competition for resources, territorial claims, and sometimes, dire conditions that pre-humans confronted in their struggle for survival. Here are a few examples,

1. Prehistoric depictions show pre-human remains pierced with arrows, such as those from the Aurignacian-Périgordian era (circa 30,000 years old) and the early Magdalenian period (circa 17,000 years old). These illustrations may signify confrontations over resources, potentially involving fatal encounters with intruders. However, other interpretations, including capital punishment, human sacrifice, targeted assassinations, or organized warfare, remain plausible.

2. Archaeological records unveil a prehistoric tragedy at Jebel Sahaba, where a population linked to the Qadan culture in far northern Sudan suffered a massacre. The site contains well-preserved human remains dating back approximately 13,000 to 14,000 years, with nearly half of the skeletons bearing embedded arrowheads—a grim indication of potential warfare casualties.

3. Several scientists have proposed hypotheses suggesting that genocidal violence could have contributed to the extinction of the Neanderthals.

4. Human remains dating from 13,500 to 23,500 years ago, originating from regions across northern and western Europe, exhibit evidence of cannibalism. These remnants share a common cultural context known as the Magdalenian, implying that the consumption of the deceased was a shared practice during this era.

5. Gough's Cave, an eminent archaeological site in southeastern England, is renowned for the discovery of 15,000-year-old pre-human skulls believed to have been fashioned into cups. Additionally, the cave contains bones that exhibit markings indicative of gnawing by other pre-human individuals.

Scientific findings and hypotheses strongly suggest that the angels indeed bore witness to the violence and bloodshed perpetrated by the pre-human inhabitants of the Earth at the time when they expressed their concerns. In response, God asserted His superior knowledge, to which the angels ultimately conceded, acknowledging that their understanding is limited to what is divinely granted to them by the All-Knowing God.

Chapter Eight: Adam

In both Aramaic and Hebrew, the name "Adam" is closely related to the concept of being formed from the earth or ground. The Hebrew etymology of the name "Adam" is commonly linked to the word "adamah," which also means "earth" or "ground."

Adam is commonly understood as the first human being among Muslims. However, it's noteworthy that the Quran does not explicitly refer to him as the first human being. As discussed in previous chapters of this book, Bashar, upon reaching a certain developmental stage, undergoes a transformative process. With the infusion of a Rooh (Spirit), Bashar evolves into a fully realized human being.

It is plausible that the entire population of the Bashar species experienced the spiritual infusion collectively, and the term "Adam" may serve more as a descriptor or attribute rather than a designated name for an individual, possibly representing the entire species. Furthermore, it is conceivable that God chose Adam, a person, to act as the representative of these newly formed human beings, instructing the angels to bow to him and subjecting him to a test along with his wife, as mentioned in the Quran.

Story of Adam in the Quran

Refer to Group 7, Adam, of Chapter Four: "Extraction of Verses." The expansive narrative of Adam and Eve in the Quran offers a rich tapestry of theological and moral insights, worthy of a comprehensive exploration. However, to focus on the primary topic of Evolution, only a condensed overview of key elements in the story is provided, avoiding unnecessary diversion.

Here are some points commonly interpreted by Muslim scholars and widely recognized by the public that are derived from a combination of accurate and potentially inaccurate information. These interpretations are drawn from the Quran, as well as references in the Bible and Jewish literature, providing a blend of perspectives on the

The Quranic narrative on Human Evolution

story of Adam. It is crucial to approach these interpretations with a discerning eye, acknowledging the variations and potential biases inherent in religious traditions and cultural narratives,

1. Adam is regarded as the first human being and, consequently, considered as the father of all humankind.

2. Adam had no parents; he was created by God from the soil of the earth, and a spirit was inserted into his body, transforming him into a human being.

3. Eve was created from one of Adam's ribs.

4. The entire process of creation unfolded in Paradise.

5. Adam and Eve were placed in Paradise for a test.

6. Upon being deceived by Satan in Paradise and consuming the forbidden fruit, Adam and Eve were punished by being sent to Earth.

7. Initially, Adam and Eve were the only inhabitants on Earth until they began reproducing.

8. The process of reproduction started with Eve giving birth to twins—first a boy and a girl in the morning and then another set of twins in the evening.

9. The morning-born boy married the evening-born girl, and vice versa, to establish the population on Earth.

10. This arrangement was temporarily permitted by God and was not considered incest.

In exploring the Quranic perspective on Adam, it becomes evident that the traditional story may not be an exact match. The attempt here is to unravel the authentic essence of the narrative as presented in the Quran. This involves delving into the scripture without imposing predetermined beliefs and being open to a deeper understanding that harmonizes with both religious teachings and scientific principles.

Adam – Father of Humanity

The Quran does not explicitly declare Adam as the first human being but associates all of humanity under a collective identity of "Bani Adam."– "بَنِي آدَمَ" in Arabic. The Quranic term Bani Adam, often translated as "Children of Adam" or "Descendants of Adam," is indeed a significant concept that carries deep theological and intellectual implications. While it is commonly understood to refer Adam as the sole biological progenitor, it is essential to explore the broader connotations of this term,

In the Quran, the use of "Bani Adam" serves to establish a collective identity for all human beings. Instead of solely emphasizing direct biological lineage, it underscores the shared human experience. Adam and Eve, being the first of the human beings, are undoubtedly parent figures as Chapter 7 Verse 27 of the Quran addresses all humankind and states, *"O Bani Adam! Let not Satan tempt you as he removed your parents from the garden, stripping them of their clothing to show them their private parts."* Here, the term "parents" is used for Adam and Eve. This usage is not contradictory, similar to Chapter 33 Verse 6, which asserts, *"The Prophet (Muhammad, peace be upon Him) is more worthy of the believers than themselves, and his wives are their mothers."* Just as this statement doesn't imply that Prophet Muhammad's wives are the biological mothers of Muslims, the reference to Adam and Eve as "parents" also does not necessarily denote biological parentage.

Quran employs specific terms, such as "Aal" – "آل" in Arabic, to refer to biological descendants or family lines. For example, "Aal e Ibrahim" signifies the biological lineage of Abraham, while "Aal e Imran" relates to the family of Imran. These terms are used when emphasizing a direct genealogical connection. In contrast, the term "Bani" is employed to signify a tribe or a clan. This convention is observable in the Quran, where "Bani Israel" is repeatedly mentioned, referring to the Children of Israel. During the era of Prophet Muhammad (peace be upon Him), there were numerous tribes and

122

clans with names beginning with "Bani." Examples include Bani Umayyah, Bani Thaqif, and Bani Uqayl.

In Arab tradition, the practice of naming tribes or clans after influential or significant figures within the community is a widespread custom. This tradition is rooted in the intention to honor and perpetuate the memory of individuals who have left a lasting impact on the community. A prominent illustration of this naming convention is evident in the clan of Prophet Muhammad (peace be upon Him), known as "Bani Hashim." Despite being the son of Abd Manaf, Hashim, the great-grandfather of Prophet Muhammad (peace be upon Him), takes precedence in the clan's name, reflecting his noteworthy contributions and prominence.

Maulana Modudi, in his influential work "Tafheem ul Quran," highlights a compelling notion. He expounds on the idea that individuals, though not biologically descended from Israel (Prophet Yaqub or Jacob), gradually became incorporated into the community known as Bani Israel, or the children of Israel. This inclusion, he underscores, unfolded over time, emphasizing the remarkable assimilation of those who accepted the faith upon the invitation of Israel or other prophets. Similarly, the use of "Bani Adam" in the Quran serves to establish a collective identity for all humankind, associating them with the very first human being.

In light of these considerations, the use of Bani Adam in the Quran does not explicitly designate Adam as the sole biological progenitor of the entire human race. Instead, it recognizes Adam as an ancestral father figure. It underscores the shared human heritage and universal brotherhood among all individuals. This concept allows for a more inclusive and transcendent understanding of human unity that extends beyond biological lineage. For instance, according to Islamic belief, Prophet Jesus had no human father; His birth was miraculous, and He is still considered part of Bani Adam despite lacking a paternal lineage. The acknowledgement of Jesus as part of Bani Adam emphasizes the broader inclusivity of the term, extending beyond conventional familial ties.

To interpret Bani Adam as specifically denoting the biological children of Adam, one would need to consider the name Adam as an attribute or set of traits shared by more than one individual. This interpretation implies that the term Adam is not limited to a single progenitor but represents a broader group of people who collectively embody the same characteristics or attributes (formed from Earth, as in Aramaic and Hebrew). However, this interpretation might be a stretch, especially considering the clear signs and context provided in the Quran about a person named Adam.

Nevertheless, two prominent figures in Islam have also made statements related to this topic. Imam Muhammad Al-Baqir, from Ehl ul Bait (family of prophet Muhammad), asserted, "Before Adam who is our father, a million Adams or more had passed from existence". Similarly, Imam Jafar Al-Sadiq, the son of Imam Baqir, expressed, "Perhaps you think that God never created a human being other than yourselves. Nay, but God created a million Adams and you are the last of those Adams". These statements offer intriguing perspectives that can prompt further contemplation on the concept of Adam. The idea that there were numerous Adams before the acknowledged father of humanity, and the notion that humans are the last in a series of creations, invite reflection on the broader narrative of human existence.

In the Quran, the existence of a historical figure named Adam is explicitly mentioned as the first human and prophet. Furthermore, the Quran details the specific people who are direct biological descendants of this Adam. In Chapter 19 Verse 58, the Quran articulates, *"Those are the ones upon whom God has bestowed favor from among the prophets of the descendants of Adam and of those We carried [in the ship] with Noah, and of the descendants of Abraham and Israel, and of those whom We guided and chose. When the verses of the Most Merciful were recited to them, they fell in prostration and weeping."*

It's noteworthy that the Arabic term used in the verse is "Zuriyati Adam," (ذُرِّيَّةِ آدَمَ) translating to "the descendants or progeny of Adam,"

The Quranic narrative on Human Evolution

instead of the most commonly used term "Bani Adam." This profound choice of wording emphasizes the direct biological lineage of Adam distinct to, what is generally understood, all humankind.

The verse intricately weaves together different lineages, encompassing the descendants of Adam, those with Noah, and the lineage of Abraham and Israel. This mosaic illustrates the diversity among communities and peoples who, spanning generations, were recipients of divine guidance through various prophets. It's worth considering that the lineage highlighted in the verse may indeed follow a biological trajectory, evidenced by the sequential descent from Adam to Noah and subsequently to Abraham, with his descendants like Isaac and Israel.

Moreover, a compelling proposition emerges suggesting that the names of all prophets mentioned in the Quran could potentially constitute a biological lineage tracing back to Adam. However, substantiating this claim necessitates a dedicated research effort. Investigating the genealogical connections between the prophets named in the Quran might unveil a cohesive biological thread linking them to, the first prophet and human, Adam. This hypothesis opens a pathway for further exploration and scholarly inquiry to establish the intricate familial ties among the esteemed prophets referenced in the Quran.

Scientific revelations on this topic are intriguing. Homo sapiens is believed to have appeared in Africa approximately 300,000 years ago. As time progressed, diverse groups of Homo sapiens embarked on migrations out of Africa in successive waves. The predominant scientific model, known as the Out-of-Africa hypothesis, posits that Homo sapiens originated in Africa and subsequently dispersed to inhabit other regions of the world. While migration routes varied, a common trajectory involved moving through the Sinai Peninsula, into the Middle East, and further into Asia and beyond. Coastal migrations also played a role in the movement of Homo sapiens. These migrations ultimately resulted in the colonization of various continents, including Asia, Europe, Australia, and the Americas.

The Quranic narrative on Human Evolution

Genetic studies, especially those focusing on mitochondrial DNA and the Y-chromosome, have provided valuable insights. Genetic diversity decreases with distance from Africa, supporting the idea of a single origin. Archaeological evidence, such as tools and fossils, supports the migration patterns. Fossils of early Homo sapiens have been found in different parts of the world, indicating their presence outside Africa.

During their migration, Homo sapiens encountered and, in some cases, interbred with other hominin species such as Neanderthals and Denisovans.

Neanderthals and Denisovans were distinct hominin species. Neanderthals occupied regions in Europe and Asia prior to the arrival of Homo sapiens. On the other hand, Denisovans are known from genetic evidence extracted from a finger bone discovered in the Denisova Cave in Siberia.

Genetic studies suggest interbreeding occurred between Homo sapiens and both Neanderthals and Denisovans when these groups came into contact. As a result, there is evidence of shared Neanderthal DNA in the genetic makeup of modern non-African human populations. Additionally, some contemporary human populations, particularly in Asia and Oceania, carry Denisovan DNA, indicating interbreeding events in the past.

Interestingly, indigenous tribes residing outside of Africa also exhibit a distinct genetic makeup attributed to pure Homo sapiens DNA. Among these tribes are those found in the Andaman and Nicobar Islands, located in the Bay of Bengal, which are home to both contacted and uncontacted indigenous groups. Uncontacted tribes, such as the Sentinalese, maintain a deliberate and voluntary isolation, limiting their interactions with the broader world.

Genetic analyses have revealed that the Andamanese, including the Sentinalese, trace their ancestry to population who embarked on a migration out of Africa approximately 50,000 to 70,000 years ago. This pioneering population navigated across the tropical expanse of

The Quranic narrative on Human Evolution

the Indian Ocean, eventually reaching and settling in the Andaman and Nicobar Islands around 45,000 to 50,000 years ago, during the broader dispersal to Southeast Asia and Australia.

This examination aims to illuminate the presence of diverse ancestral DNA within modern humans, challenging the notion that Adam and Eve were the exclusive biological progenitors of humanity. The genetic composition of contemporary humans includes strands associated with Homo sapiens, Neanderthals—thought to have gone extinct approximately 40,000 years ago and Denisovans—whose existence persisted until at least 50,000 years ago. The coexistence of these distinct genetic lineages in modern populations underscores the intricate tapestry of human ancestry, offering a nuanced perspective that goes beyond the premise of a singular ancestral pair.

The term "scientific Adam and Eve" is often used to describe a hypothetical pair of human ancestors from whom all modern humans are believed to have descended. This concept is rooted in genetic studies that aim to trace the ancestry of the human population by analyzing DNA.

In scientific terms, the most recent common ancestor (MRCA) is the individual from whom all living humans are directly descended. The use of the term Y-chromosomal Adam and Mitochondrial Eve in this context is a metaphorical reference to these common ancestors. The scientific Adam and Eve are not envisioned as a single couple but rather as two individuals who lived at different times in the distant past.

Y-Chromosomal Adam: It refers to the most recent common ancestor of the patrilineal line of all currently living human males. It is a concept in human genetics that traces the lineage of the Y chromosome, which is passed from father to son.

Estimates for the existence of Y-chromosomal Adam vary, but most genetic studies suggest that he lived between 100,000 to 200,000 years ago. The concept of Y-chromosomal Adam is based on population genetics and the study of genetic markers on the Y chromosome.

The Quranic narrative on Human Evolution

Geneticists use these markers to trace lineages and estimate when certain common ancestors might have lived. Understanding the genetic markers on the Y chromosome helps scientists reconstruct human migration patterns and evolutionary history. It provides insights into the movement of populations and the divergence of human groups over time.

The concept has limitations, and it doesn't imply that Y-chromosomal Adam was the only man alive during his time. Other males from the same period may have descendants as well, but their patrilineal lines didn't survive to the present. "Y-chromosomal Adam" is a symbolic title and should not be confused with religious or mythological figures. It represents a scientific estimate based on genetic data and the study of human evolution.

Mitochondrial Eve: It is a concept in human genetics that refers to the most recent common ancestor of the matrilineal (female-line) ancestry of all living humans. The concept is based on the study of mitochondrial DNA, a type of DNA found in the mitochondria, structures within cells. Unlike nuclear DNA, which is a combination of genetic material from both parents, mitochondrial DNA is inherited exclusively from the mother.

Estimates for the existence of Mitochondrial Eve vary, but most genetic studies suggest that she lived between 100,000 to 200,000 years ago. Geneticists use mitochondrial DNA markers to trace maternal lineages and estimate when certain common ancestors might have lived. By studying the mutations in mitochondrial DNA, researchers can reconstruct the evolutionary history of human populations.

The concept has limitations, and it doesn't imply that Mitochondrial Eve was the only woman alive during her time. Other females from the same period may have descendants as well, but their matrilineal lines didn't survive to the present. Y-chromosomal Adam and Mitochondrial Eve are often discussed in conjunction, but it's important to note that these individuals did not live at the same time or in the same place.

Parents or No parents?

As it is already discussed that the narrative of human existence unfolds through a series of transformative stages, commencing with the emergence of life from mud, evolving into cellular structures, and progressing to a more complex phase of asexual reproduction. Over time, this biological evolution transitioned into the intricate process of sexual reproduction. Within this evolutionary journey, a pivotal point is reached with the emergence of a Bashar or pre-human being. This entity undergoes a profound transformation, culminating in the creation of a complete human being, Adam, marked by the infusion of the Rooh or spirit.

The prevailing perspective among Muslims often revolves around the notion of Adam's special creation, an idea widely acknowledged and accepted within the Islamic discourse. This understanding posits that Adam's formation was a unique occurrence, distinct from the natural processes or involving a parental lineage. Even theologians who entertain the idea of evolution often carve out an exception for Adam, contending that while the broader evolutionary processes might hold true, the genesis of Adam was a specialized event.

The solitary verse in the Quran that is often cited as supporting the idea of Adam having no parents is found in Chapter 3 Verse 59. It states, *"Indeed, the example of Jesus to God is like that of Adam. He created Him from dust; then He said to him, "Be," and he was,"* implying that Adam and Jesus, were created by both types of divine creation, Alam-e-Khalq (the World of Creation) and Alam-e-Amr (the World of Command). However, the commonly perceived implication of this verse is that Adam had no parents, as the Quran draws a parallel between Adam and Jesus. Similar to Jesus, who was born miraculously without a father, the suggestion is that Adam's origin was also unique in its divine creation.

If the Quran had stated the opposite, such as "The example of Adam is like that of Jesus," the inference of Adam having no father might have been more straightforward. Alternatively, a clearer statement such as, "Jesus had no father, just as Adam had neither father nor mother," would have removed ambiguity. However, the interpretation of this verse is often influenced by preconceived notions. The context of the verse primarily revolves around Jesus and aims to counter the claims of those who attribute divinity to him based on his miraculous birth without a father.

In this verse, Jesus is likened to Adam in the context of their creation. Adam's formation involved originated from mud or dust, mentioned in the Quran as Nafsin Wahid, or a single cell that ultimately evolved to the stage of a Bashar (a pre-human being). Subsequently, God's command, "be", was given to infuse him with a spirit, transforming him into a complete human being. Similarly, the creation of Jesus is portrayed as a divine act where the absence of a human father is compensated by God's command, "be," bringing him into existence as a human being. This comparison underscores a commonality in the creative process, highlighting the role of divine command in both instances and not suggesting that Adam had no parents.

Furthermore, Muslim scholars, perhaps inadvertently, have caused confusion by equating the verses describing the origin of life through the single cell, Nafsin Wahida, with the creation of Adam. Hence, understanding the creation of Adam without the process of evolution or involvement of parents.

Moreover, they have linked the process of asexual reproduction of this single cell with the creation of Eve from Adam, or more specifically, with the concept of Eve being created from Adam's rib. This confusion arises from the misinterpretation of a Hadith, a saying of Prophet Muhammad (Sahi Bukhari-5184), in which he metaphorically equated the inflexibility of women with a rib.

The Quran emphasizes twice that Adam was chosen, suggesting the existence of other options, out of which Adam was selected. He was chosen to be the first human being and to be a prophet of God. Chapter 3 Verse 33 states, *"Indeed, God chose Adam and Noah and the family of Abraham and the family of 'Imran over the worlds."* and Chapter 20 Verse 122 states, *"Then his Lord chose him, and turned towards him with forgiveness, and granted him guidance."*

These two verses imply the existence of other individuals alongside Adam and Eve, with Adam being selected from among them. Hence, there existed a population, and Adam and Eve, being part of it, likely had parents. This aligns seamlessly with our examination of evolution in the context of the Quran and scientific discoveries.

Place of Creation and Test of Adam

Chapter 38 Verse 72 of the Quran mentions the order of God to the angels when Adam was being transformed to the first human being. It states, *"So, when I have proportioned him and breathed into him of My [created] soul, then fall down to him in prostration."*

Chapter 2, Verse 34 of the Quran encompasses the subsequent event, *"We (God) said to the angels, "Prostrate to Adam," and they prostrated, except for Iblis (Satan)."* The angels bowed to Adam, but Satan refused, leading to his expulsion by God. Satan's identity as a djinn is mentioned in Quran Chapter 18 Verse 50, *"He was one of the djinns, so he chose the way of disobedience to his Lord's Command."*

Following this, Adam and Eve, were placed in a garden for a test. The purpose was to illustrate that Satan is their adversary, and he would perpetually try to deceive them, leading them away from the righteous path. Chapter 7 Verse 19 of the Quran states, *"And [O Adam], dwell, you and your wife, in the garden and eat from wherever you will but do not approach this tree, lest you be among the wrongdoers."*

Adam and Eve were granted the freedom to partake in all the pleasures of the garden, with the exception of eating the fruit of one specific tree. Despite this clear directive, Satan cunningly manipulated them, leading them to consume the forbidden fruit. Chapter 7 Verse 20 encompasses, *"He (Satan) said (to Adam and Eve), Your Lord did not forbid you this tree except that you become angels or become of the immortal."*

The Quranic narrative on Human Evolution

As a consequence of Satan's deceptive influence, Adam and Eve succumbed to temptation, leading to their disobedience. In response, God instructs them to depart from the garden. Chapter 2 Verse 36 states, *"So Satan caused them to slip out of it (the garden) and removed them from that [condition] in which they had been. And We said, 'Go down, [all of you], as enemies to one another, and you will have upon the earth a place of settlement and provision for a time."*

Adam acknowledges his mistake and repents to God, who accepts his repentance. Despite this, God instructs them to leave the garden and settle down. Additionally, God advises them to follow His guidance when bestowed upon them. Chapter 2 Verse 38 states, *"We (God) said, 'Go down from it (the garden), all of you. And when guidance comes to you from Me, whoever follows My guidance, there will be no fear concerning them, nor will they grieve."*

The prevailing view among numerous Muslim scholars and the followers is that the creation of Adam, along with the subsequent test involving Adam and Eve, took place in paradise. According to this perspective, following the disobedience of Adam and Eve, God instructed them to depart from paradise and reside on Earth. Nevertheless, it's vital to note that this interpretation diverges from the narrative presented in the Quran.

Two terms have fueled the belief that these events unfolded in paradise. One of these terms is the Arabic word "Jannah" (الْجَنَّةَ), often translated as "paradise" or "garden." The utilization of "Jannah" is evident in Chapter 7 Verse 19 which states, *"And [O Adam], dwell, you and your wife, in the Jannah (paradise/garden) and eat from wherever you will but do not approach this tree, lest you be among the wrongdoers"*. This was indeed a distinct place where the food was in abundance, also, Chapter 20 Verse 119 showing that there was an abundance of water and shade available, *"And indeed, you will not be thirsty therein or be hot from the sun."*

The other term contributing to the belief that these events occurred in paradise is the use of the term 'Ahbitu' (اهْبِطُوا) or 'Ahbita' (اهْبِطَ),

The Quranic narrative on Human Evolution

translated as "go down" or "descend." This term is commonly interpreted as signifying the descent of Adam and Eve from Paradise to Earth. For instance, Chapter 7 Verse 24 states, *"Go down, being enemies to one another. And for you on the earth is a place of settlement and enjoyment for a time."*

By mistranslating these two terms, the entire narrative of the Quran becomes misconstrued.

Jannah: Paradise or Garden? The Arabic word "Jannah" (الْجَنَّةُ) is indeed used in the Quran to refer to paradise, the ultimate reward for the righteous in the hereafter, where residents will stay forever. However, it's important to note that the word "Jannah" (الْجَنَّةُ) is also used in the Quran to denote an earthly garden. In fact, there are more than 15 places in the Quran where the word "Jannah" is used to describe as an earthly garden. For example,

Chapter 6 Verse 99 encompasses, *"And [We produce] gardens (جَنَّٰتٍ) of grapevines and olives and pomegranates, similar yet varied. Look at [each of] its fruit when it yields and [at] its ripening."*

Chapter 6 Verse 141 encompasses, *"And He it is who causes gardens (جَنَّٰتٍ) to grow, [both] trellised and untrellised, and palm trees and crops of different [kinds of] food and olives and pomegranates, similar and dissimilar."*

Chapter 17 Verse 91 states, *"Or [until] you have a garden (جَنَّةٌ) of palm trees and grapes and make rivers gush forth within them in force [and abundance]."*

Chapter 23 Verse 19 states, *"And We brought forth for you thereby gardens (جَنَّٰتٍ) of palm trees and grapevines in which for you are abundant fruits and from which you eat."*

Chapter 18 Verse 35 encompasses, *"And he entered his garden (جَنَّتَهُ) while he was unjust to himself."*

Chapter 36 Verse 34 states, *"And We placed therein gardens (جَنَّٰتٍ) of palm trees and grapevines and caused to burst forth therefrom some springs."*

Adam and Eve were residing in a garden on Earth, where there was an abundance of food, water, and shade to protect them from the weather. This garden should not be confused with the ultimate paradise promised for the righteous after the Day of Judgment, where the rewarded individuals will dwell forever without the risk of being expelled. The Quran states in Chapter 3 Verse 198, *"But those who feared their Lord will have gardens (جَنَّٰتٌ) beneath which rivers flow, abiding <u>eternally</u> therein, as accommodation from God. And that which is with God is best for the righteous."*

Ahbitu: Go Down or Settle Down? The terms "Ahbitu" (اهْبِطُوا) and "Ahbita" (اهْبِطْ) are translated as "Go down" or "Descend" in Chapter 7 Verse 24, and Chapter 20 Verse 123 of the Quran. However, the same word is used in another verse of the Quran, mentioning the time when the Children of Israel (Bani Israel) told Prophet Moses that they could not endure one kind of food, Man-o-Salwa (also known as Manna and Quail).

Chapter 2 Verse 61 recounts the interaction between Moses and the Bani Israel, *"He (Moses) said, What! You wish to exchange the better for something inferior? Therefore, settle down (اهْبِطُوا) in Egypt or any city, where you will get what you demand."* The majority of scholars have translated the Arabic word (اهْبِطُوا) as "go down in settlement," "go down to a town," or "settle down in Egypt." Notably, none of these translations suggests a descent from a high place or from paradise. Instead, they emphasize a change in location or situation, aligning with the broader understanding that the term is context-dependent and denotes a shift or change in circumstances rather than a physical descent from a celestial realm.

Remarkably, the circumstances of Bani Israel at that time mirrored those of Adam and Eve in the garden. The Quran recounts a period when Bani Israel found themselves in the Tih desert after Prophet Moses liberated them from the Pharaoh of Egypt. Chapter 2 Verse 57 includes, *"And We shaded you with clouds and sent down to you manna and quails, [saying], "Eat from the good things with which We have provided you."*

Chapter 2 Verse 60 reads, *"And [recall] when Moses prayed for water for his people, so We said, 'Strike with your staff the stone.' And there gushed forth from it twelve springs, and every person knew its watering place. 'Eat and drink from the provision of God, and do not commit abuse on the earth, spreading corruption."*

In both, the situations of Bani Israel in the Tih desert and the scenario of Adam and Eve in the garden, God provided essential elements for sustaining life such as abundant food, water, and shade from the sun. The reference to ample provisions signifies that God ensured the well-being of the recipients by supplying everything necessary for their sustenance and comfort. Similarly, in both situations, God directed the beneficiaries to leave this state of blessing and settle down where they make efforts to fulfil these necessities for their lives.

This refined understanding challenges the interpretation that suggests an exclusive link between 'Ahbitu' (اهبِطُوا) and a descent from Paradise to the Earth. The broader context of its usage underscores its versatile nature, emphasizing the significance of considering the specific circumstances in which it is employed. In the narrative of Adam and Eve, it aligns with the understanding that they were instructed to leave the garden, where they were enjoying life's necessities in abundance as a divine blessing. Instead of a literal descent from a celestial paradise, it suggests a directive to establish a residence on Earth, where they must work to attain these provisions, shifting from a state of divine bestowal to a more earthly existence.

The translation of verses with the term Ahbitu (اهبِطُوا) or Ahbita (اهبِطَ), aligns more coherently with the interpretation of (اهبِطُوا) as "settle down",

The Quranic narrative on Human Evolution

Chapter 2 Verse 36: *"So Satan caused them to slip out of it (the garden) and removed them from that [condition] in which they had been. And We (God) said, 'Go settle down, [all of you], some of you the enemies of others; and you will have upon the earth a place of settlement and provision for a time."*

Chapter 2 Verse 38: *"We (God) said, Go (from the garden) and settle down, all of you. And when guidance comes to you from Me, whoever follows My guidance, there will be no fear concerning them, nor will they grieve."*

Chapter 7 Verse 24: *"He (God) said, "Go settle down, some of you enemies of some; and for you on the earth there will be a dwelling place and enjoyment for a time."*

Chapter 20 Verse 123: *"He (God) said, "Go from it (the garden and settle down), all of you. And when guidance comes to you from Me, whoever follows My guidance, there will be no fear concerning them, nor will they grieve."*

In these cases, the surrounding context reinforces the idea that the term Ahbitu (اهْبِطُوا) or Ahbita (اهْبِطَ) is best interpreted as a directive to relocate and establish a settlement on Earth. This interpretation emphasizes a shift in location rather than a descent from a higher realm or Paradise. Therefore, it suggests that Adam and Eve were initially in a garden on Earth and were instructed to depart from the garden, to get settle down elsewhere on Earth.

Moreover, other clear verses are indicating that the unfolding events related to Adam occurred specifically on Earth rather than in paradise. The angels were informed about the creation of a being to be appointed as the Khalifa or deputy of God, specifically on Earth, as Chapter 2 Verse 30 of the Quran encompasses, ***"And [mention, O Muhammad], when your Lord said to the angels, "Indeed, I will make upon the <u>earth</u> a successive authority."*** The Quran is unequivocal in asserting that the creation of humans, intended to be a Khalifa or deputy of God, was announced to take place on Earth, rather than in the celestial realm or paradise.

Chapter 15 Verse 39 also provides a crucial perspective to prove the point. It states, *[Iblees] said, "My Lord, because You have put me in error, I will surely make disobedience attractive to them on Earth, and I will mislead them all."* In this verse, following Satan's refusal to bow to Adam, he addresses God, articulating his determination to lead humanity astray on Earth. It is noteworthy that this dialogue transpires even before Adam and Eve are placed in the garden for their test. Satan explicitly threatens to make disobedience appealing to humans, more specifically, on Earth.

With the elucidation provided by these explicit verses, there is little room for ambiguity regarding the creation of Adam and the location of the garden where the test of Adam and Eve transpired. The discernible indication is that both the creation of Adam and the testing ground were situated on Earth, distinct from the notion of a heavenly paradise.

Others Coexisted with Adam and Eve?

In the Quran, there are specific verses that unequivocally depict Adam and Eve as not being a solitary couple but accompanied by others. The emphasis on Adam being chosen, as examined previously under the heading of "Parents or no parents", adds depth to the narrative, challenging the notion of their solitary existence. In the Quran, Chapter 3 Verse 33 states, *"Indeed, God chose Adam and Noah and the family of Abraham and the family of 'Imran over the worlds."* and Chapter 20 Verse 122 states, *"Then his Lord chose him, and turned towards him with forgiveness, and granted him guidance."*

As previously explored, two possibilities exist regarding Adam's selection, either to be the first human being or to be a prophet, or maybe both as it came up twice in the Quran. However, what remains certain is the presence of other individuals, specifically Bashars, the pre-humans, during the pivotal moments when Adam was chosen to be the first human being. Additionally, there were other human beings present when Adam was chosen to assume the role of a prophet.

The Quranic narrative on Human Evolution

In the Arabic language, grammatical forms serve to convey the number of individuals being addressed. The singular form is utilized when addressing a single person, the dual form for two individuals, and the plural form for a group exceeding two. Within the Quranic depiction of Adam, there are instances where the plural form is employed, signaling that the address is directed towards a collective rather than a single or two persons. This deliberate use of the plural adds a nuanced layer to the comprehension of Adam's narrative, hinting at the involvement or presence of a community rather than a solitary couple. For example,

Chapter 20 Verse 123: *He (God) said, "Go down from it, <u>all of you</u>. And when guidance comes to you from Me, whoever follows My guidance, there will be no fear concerning them, nor will they grieve."*

Chapter 15 Verse 39: *[Iblees] said, "My Lord, because You have put me in error, I will surely make [disobedience] <u>attractive to them</u> on earth, and I will mislead them all."*

Chapter 7 Verse 16: *[Iblees] said, "Because You have put me in error, I will surely sit in <u>wait for them</u> on Your straight path."*

Chapter 7 Verse 17: *[Iblees continues] Then I will <u>come to them</u> from <u>before them</u> and from <u>behind them</u> and on their right and on their left, and You will not find <u>most of them</u> grateful [to You].*

Chapter 7 Verse 24: *He said, "Go, settle down, <u>some of you enemies of some</u>; and for you on the earth there will be a dwelling place and enjoyment for a time."*

In the passages recounting Satan's refusal to bow to Adam and the subsequent expulsion of Adam and Eve from the garden upon their failure in the test, there is a distinct indication that Adam and Eve were not the sole inhabitants in that setting. Although the focus is primarily on Adam and Eve undergoing the trial, Chapter 20 Verse 123 strongly implies the presence of others who were similarly instructed to depart from the garden.

Moreover, Chapter 7 Verse 24 of the Quran declares, "some of you will be enemies of some," directly pertaining to human beings, with Satan already established as humanity's ultimate adversary. This proclamation reinforces the notion that the directive to leave the garden and establish a dwelling was intended for the entire community cohabiting with Adam and Eve.

Indeed, the dialogue between God and Satan in the mentioned verses highlights a noteworthy point. Satan (Iblees), in his dialogue, consistently alludes to the presence of a wider community beyond just Adam and Eve. Rather than directing his words solely towards Adam or both Adam and Eve, Satan's threats and attempts to mislead indicate a broader population.

Some Muslim scholars suggest that using the plural form of grammar asserts that it encompasses Adam, Eve, and Satan. To challenge this viewpoint, one may examine the verses preceding and following Adam's repentance within the same incidents,

Chapter 2 Verse 36: *So, Satan caused them to slip out of it (the garden) and removed them from that [condition] in which they had been. And We (God) said, "Go settle down, [all of you], some of you the enemies of others; and you will have upon the earth a place of settlement and provision for a time."*

Chapter 2 Verse 37: *"So Adam received words from his Lord, and He relented toward him. Indeed, He is the Most-Relenting, Most-Merciful."*

Chapter 2 Verse 38: *"We (God) said, Go (from the garden) and settle down, all of you. And when guidance comes to you from Me, whoever follows My guidance, there will be no fear concerning them, nor will they grieve."*

The initial verse, Chapter 2 Verse 36, sets the scene before the repentance of Adam. The Quran articulates God's command, instructing them, in a punitive measure, to depart from the garden in the plural form. At this juncture, there may be uncertainty regarding the presence of Satan.

The Quranic narrative on Human Evolution

Following this, Chapter 2 Verse 37, illustrates that God pardoned them upon Adam's plea for forgiveness. Subsequently, Chapter 2 Verse 38, occurring after repentance and forgiveness, sees God once again instructing their collective departure from the garden. However, this time, it is not punitive but rather part of a divine plan.

Without a shadow of a doubt, Satan was conspicuously absent in that final scenario. The divine directive for Adam and Eve to await guidance patiently, subsequently acting upon it for success, solidifies this reality. Unquestionably, it is evident that Satan neither obtained forgiveness nor was anticipated to conform to divine guidance. The guidance was anticipated to be received and acted upon by the human beings accompanying Adam and Eve.

Another intriguing aspect is the Quran's declaration of Adam and Eve as husband and wife, found in Chapter 2 Verse 35, *"And We said, O Adam, dwell, you and your wife, in Paradise and eat therefrom in [ease and] abundance from wherever you will. But do not approach this tree, lest you be among the wrongdoers".* However, upon succumbing to temptation and eating the forbidden fruit, as mentioned in Chapter 20 Verse 121, their clothing was stripped away, *"And they [i.e., Adam and his wife] ate of it, and their private parts became apparent to them, and they began to fasten over themselves from the leaves of Paradise. And Adam disobeyed his Lord and erred."* This raises the question of whom they felt ashamed before, given their marital status. The Quran, notably, emphasizes the intimate relationship between spouses in the beginning part of Chapter 2 Verse 187, *"They are a clothing for you and you are a clothing for them".* One conceivable explanation is that their sense of shame arose from the presence of other beings cohabiting the garden with them.

Furthermore, this understanding of co-existence allows us to unequivocally dismiss the idea suggesting that initial reproduction involved Eve giving birth to sets of twins—one in the morning comprising a boy and a girl, and another in the evening—with the morning-born boy marrying the evening-born girl. This concept, lacking substantiation from either the Quran or Hadith (the sayings of

Prophet Muhammad), seems to be a product of imaginative speculation. It is crucial to underscore that the proliferation of the Earth's population did not rely on incest, as there were other human beings present on Earth.

The Quran makes specific reference to the two sons of Adam, although their names are not explicitly provided. The names Habil (Abel) and Qabil (Cain) are identifiable in the Book of Genesis within the Old Testament of the Bible. However, the widely recognized family tree implies the existence of a third son, Shees (Seth). While the Quran does not explicitly mention this additional detail about Seth, it is found in other historical and religious sources, such as the works of "Seerat Ibn Hisham" and "Tarikh al-Tabari."

Era of Adam

Determining the precise era of Adam poses a considerable challenge, marked by diverse opinions, particularly in the wake of fossil discoveries. Theologians engage in ongoing debates, seeking alignment between the theological narrative and scientific findings. The discovery of Neanderthal or Denisovan DNA within modern humans prompts reconsideration, sparking discussions about whether Adam's era predates the extinction of these hominid species.

The notion of Adam's estimated era, around 6,000 years ago, is frequently linked to a literal interpretation of specific genealogies detailed in the Bible. This approach involves adding up ages and timelines provided in biblical passages, notably in Genesis. However, it's crucial to note that the Quran remains relatively silent on this specific topic, leaving room for interpretation and discussion within Islamic theological circles. The interplay between religious narratives and scientific discoveries continues to shape perspectives on the timeline of Adam's era.

In considering various facts and research findings, one may endeavor to estimate the era of Adam. For example, Adam was the first human being, the Homo sapiens sapiens, a modern human. Consequently, the era associated with Adam predates the discovery of distinct modern traits linked to modern human beings, such as the establishment of the earliest civilizations, the founding of cities, the advent of farming, and the initiation of animal domestication.

Among the earliest civilizations uncovered by archaeological exploration are the Sumerians, who flourished in ancient Mesopotamia, located in what is now modern-day Iraq, around 4500 BCE. Revered as one of humanity's initial civilizations, Sumer stands out for its advanced urban centers, intricate social organization, a developed system of writing, and an array of technological and cultural accomplishments.

Jericho, an ancient city nestled near the Jordan River in the West Bank, stands as one of the oldest known settlements. Its archaeological site exposes layers of human habitation stretching back to approximately 8000 BCE. Jericho is renowned for its remarkably well-preserved defensive walls, reflecting a history that spans millennia.

Göbekli Tepe, a village situated in southeastern Turkey, claims the distinction of being one of the oldest known villages, with roots extending from approximately 9600 BCE to 9500 BCE. This ancient settlement is characterized by massive stone pillars meticulously arranged in circular patterns, some soaring to heights exceeding 15 feet.

Moreover, Chapter 5 Verse 27, which recounts the story of Adam's two sons, encompasses, *"And recite to them the story of Adam's two sons in truth when they both offered a sacrifice [to God]."* In interpreting this verse, many Muslim scholars have proposed an additional detail: one of the sons, a farmer, offered the yield of his farm or garden while the other son, a shepherd, offered an animal for sacrifice. This interpretation, often found in Islamic commentary, seems to draw parallels with a similar account in the Bible, and it aligns seamlessly with the Quranic narrative. This development hints at the inception of the Neolithic Era.

The hallmark features of the Neolithic Era encompassed the advent of agriculture and the domestication of animals. Originating in the Near East, particularly in the Fertile Crescent (encompassing modern-day Iraq, Syria, Lebanon, Israel, and Jordan), the Neolithic Era witnessed the emergence of agriculture in certain regions of the Fertile Crescent as early as 10,000 BCE. The tale of Adam's sons, with one cultivating the land and the other tending to livestock, resonates with the transformative shift toward settled agricultural communities, marking a significant chapter in human history.

By synthesizing and amalgamating the available data, a cautious estimate places Adam's era within the timeframe of no more than 10,000 BCE and not preceding 9500 BCE. It is crucial to acknowledge that this determination hinges on a multitude of factors and considerations. As our understanding evolves through ongoing developments and fresh discoveries, the chronological placement of Adam's era can undergo reassessment, guided by the same foundational principles.

Reconstructed Story of Adam

Following the research done, the reconstructed story of Adam is as follows,

1. Adam and Eve were born as Bashar (pre-humans) on Earth, to parents who were also Bashars.

2. God chose Adam to be the first human being, setting him apart for a significant purpose.

3. God infused Adam with a spirit, transforming him into a full-fledged human being. Angels and Iblis (Satan) were instructed to bow down to Adam.

4. Angels followed God's command and bowed down to Adam, symbolizing their obedience. However, Satan, considering himself superior, refused to comply.

The Quranic narrative on Human Evolution

5. Adam and Eve were husband and wife and other people coexisted with them.

6. God directed Adam and Eve to reside in a garden where life's necessities abounded. They were instructed not to eat from a specific tree.

7. Satan deceived Adam and Eve into eating the forbidden fruit. Consequently, God instructed them, along with the people in their company, to leave the garden and settle on Earth.

8. Adam repented, and God, in His mercy, forgave them. Adam was then chosen as a prophet, with the announcement that divine guidance would be sent, and success would be attained by those who accepted it.

9. Adam and Eve had three sons: Abel, Cain, and Sheth (in Arabic: Habil, Kabil and Shees)

10. Adam, as the first human being, is considered the father of humanity in a symbolic and spiritual sense, rather than a direct biological ancestor.

With the revised narrative of Adam in the framework of human evolution, the research achieves its primary goal. The extensive findings not only fulfil the research's purpose, which was to scrutinize the Quran's stance on human evolution but also accomplish an additional objective—demonstrating the compatibility between the Quran and scientific principles. The subsequent section of the book provides a concise summary of the research, aiming to articulate the Quran's perspective in a succinct manner.

Chapter Nine: Summary of Research

The exploration of the Quranic narrative on human evolution has been marked by its profound nature. In Chapter 54 Verse 17, the Quran declares, *"We have made this Qur'an easy as a reminder. Is there, then, any who will take heed?"* This proclamation set the stage for unbiased research, delving into the evolution of human beings as depicted in the Quran. The research was driven by a genuine quest for divine knowledge on this subject, devoid of any preconceived notions or predetermined outcomes. Given the Quran's assertion as the word of God and the acknowledgement of God as the creator of all, including humans, the research sought logical alignment between divine precepts and practices.

Origin of Life

The Quran, in Chapter 19 Verses 9 and 67 asserts, *"We (God) created him (human) before, while he was nothing,"* implying the origin of life from non-living matter. Subsequently, life originated from a single cell, referred to as Nafsin Wahida in various Quranic verses, such as Chapter 4 Verse 1, which states, *"O mankind, fear your Lord, who created you from one soul (living organism or a biological cell)."* This single cell is elucidated as the starting point of the creation of the human being. In scientific terms, this process aligns with the Abiogenesis hypothesis, suggesting that life originated from non-living matter through a complex series of chemical reactions.

The Quran provides additional insights into the composition of this non-living matter, particularly emphasizing the role of clay. Numerous verses detail that the creation of the human being commenced with clay, suggesting that the single cell from which the human being originated was formed from this substance. Various verses encapsulate the diverse stages of the chemical reaction involving clay. For instance, Chapter 7 Verse 32 of the Quran declares, *"[God] who perfected everything which He created and began the creation of man from clay."*

The Quran introduces another crucial ingredient in the creation process, placing emphasis on water. Chapter 25 Verse 54 includes, *"And it is He (God) who has created from water a Bashar."* The integration of these ingredients to form a single living organism is a topic under scientific investigation. Scientific hypothesis further posits that the single cell evolved to give rise to distinct branches of organisms: Bacteria, Archaea, and Eukarya. According to this view, all present living things have evolved from that single cell, a concept explained by the term LUCA.

While LUCA, the Last Universal Common Ancestor, may not represent the very first single cell formed, it is believed that every form of life on our planet can trace its lineage back to LUCA which is considered the most recent common ancestor of all living organisms. Scientists identify LUCA by examining the genetic material (DNA and RNA) of various living organisms, revealing common genes and genetic sequences shared among diverse life forms. This commonality in genetic material provides insights into the shared ancestry of all living organisms.

These common genes and genetic sequences may suggest a common ancestor. However, this shared genetic information could also be interpreted as evidence of a common designer, referring to a divine creator who implemented similar genetic codes across different species. The concept is posited in Intelligent Design theory.

In the Quran, there is one specific mention of water in relation to the creation of every living thing. Chapter 21 Verse 30 encompasses, *"(God) made from water every living thing?"* Beyond water being highlighted as a common ingredient in all forms of life, the Quran does not explicitly provide other links in this context. As the Quran does not explicitly oppose such concepts, their exploration and evaluation fall within the realm of science, to be judged on their own merits.

Reproduction

The Quran articulates in various verses that Nafsin Wahida - the single cell reproduced to create its couple. For example, Chapter 7 Verse 189 and Chapter 39 Verse 6 encompass, *"He (God) created you from one soul and created from it its couple"* implying that the single cell gave rise to its counterpart. The Quranic interpretation of this stage aligns seamlessly with scientific understanding. The scientific term for this process is Mitosis, which refers to asexual reproduction wherein a cell divides into two identical cells.

Science characterizes the next stage of evolution in reproduction as Meiosis, a term used to describe the process of sexual reproduction. The transitional stages between asexual reproduction and sexual reproduction are still being explored by science. Chapter 35 Verse 11 in the Quran describes the stage of sexual reproduction, but a more profound scrutiny and examination of the verse is needed for a comprehensive understanding. The verse states, *"And God created you from dust, then from a Nutfa (sperm or reproductive cell); then He made you mates."*

The key is to analyze the chronological order.

1. Creation from dust;

2. Then, creation from Nutfa (Sperm);

3. Then, pairing up as mates.

It delineates a single cell's **creation from dust**, aligning with Abiogenesis, followed by asexual reproduction or Mitosis in scientific terms. Subsequently, the **creation from Nutfa** or productive cells or sperm, termed as Meiosis in scientific discourse, explicates the concept of sexual reproduction. The reference to the **pairing up of mates** or spouses in the latter part of the verse signifies that, as a consequence of these processes, a multitude of males and females emerge, forming pairs and engaging in reproduction.

Bashar or Homo sapiens

Bashar is a Quranic attribute that denotes the earthly, animalistic facets of human nature, encapsulating instincts and inherent inclinations. Within this framework, Human beings, referred to as "Insaan" in the Quran, can be categorized as Bashar due to the manifestation of these animalistic qualities. It's crucial to emphasize that, although human beings can be characterized as Bashar, the reciprocal statement is not valid; Bashar cannot be equated with human beings or Insaan.

Prior to the emergence of life from a single cell, God communicated to the angels about the creation of a being referred to as Bashar. Chapter 15 Verse 28 includes, *"When your lord said to the angels, I [God] am creating Bashar from clay from molded mud"* and Chapter 38 Verse 71, states, *"When your lord said to the angels, I [God] am about to create Bashar from clay."*

The Quran does not outline any intermediary phases between meiosis and the formation of Bashar. Rather, it alludes to a period in human history deemed inconsequential. Chapter 76 Verse 1 of the Quran articulates, *"Has there [not] come upon human (Insaan) a period of time when he was not a thing worth of mentioning."* This verse suggests a phase in human existence that is considered insignificant and not explicitly elaborated upon.

As the Quran does not explicitly delineate these intermediary stages, the specific details of whether our ancestors were apes or similar questions remain unanswered within the Quranic framework. The divine wisdom behind this absence of explicit information is beyond human comprehension. It's crucial to emphasize that the absence of explicit denial in the Quran implies that such concepts are not deemed "Haraam" or forbidden. Consequently, the evaluation of these ideas should be approached based on their own merits and within the scope of scientific inquiry.

Human being or Homo sapiens sapiens

In Chapter 38 Verse 72 of the Quran, God imparts information to the angels about the creation of the first human being, stating, *"So when I have proportioned him (Bashar) and breathed into him my [created] Rooh (spirit), then fall down to him in prostration."* The word "proportioned" equates to the process of Evolution, which the Quran does not detail, deeming it not worth mentioning.

This verse elucidates two distinct realms of divine creation. The first, Alam-e-Khalq - "خَلْق" (the world of creation), denotes the gradual process of bringing something into existence, involving formation and crafting that unfolds over time. In contrast, the second form of divine creation operates with immediacy, manifesting instantly upon God's command. This type of creation is referred to as Alam-e-Amr – "أَمْر" (World of Command).

The creation of Bashar followed the gradual process of Alam-e-Khalq - "خَلْق" (the world of creation). It commenced with the formation of a single cell fashioned from mud, progressing through various stages of evolution to reach the stage of Bashar. Subsequently, the infusion of a spirit occurred through Alam-e-Amr – "أَمْر" (World of Command), signifying the immediate, divine act that bestowed spiritual essence upon Bashar, completing the creation of the human being.

In this context, the initial segment of Chapter 38 Verse 75 of the Quran reads, *"The Lord said: O Iblis (Satan), what prevented you from prostrating yourself before him whom I created of My Two Hands."* The metaphor "two hands" is used symbolically, representing the creation of human beings through both forms of divine creation— Alam-e-Khalq (the World of Creation) and Alam-e-Amr (the World of Command).

This transformation, of Bashar to Human, extends beyond the physical realm; it signifies a profound metaphysical shift encompassing various crucial aspects, such as spiritual essence, consciousness, emotions, intellect, and metaphysical connection. The Rooh (spirit) plays a pivotal role in this transformation, not only elevating Bashar to the status of a human being but also designating them as God's Khalifa, a deputy on Earth. Chapter 2 Verse 30 of the Quran states, *"And [mention, O Muhammad], when your Lord said to the angels, "Indeed, I will make upon the earth a successive authority."*

This verse highlights that God has chosen human beings to serve as His Khalifa (خَلِيفَة) or Caliph, endowing them with a unique role and responsibility. As Khalifa, human beings possess distinct attributes that set them apart from various other beings, including free will to choose between good and evil, for which they will be held accountable to God. The role of Khalifa entails upholding moral and ethical principles, including justice, compassion, and fairness in all aspects of life. Additionally, human beings are expected to engage in spiritual growth, fostering a relationship with the divine through worship, prayer, and self-purification.

In the latter part of Chapter 2 Verse 30 of the Quran, the angels respond to God's announcement of placing a Khalifa on Earth, expressing concern, *"They (angels) said, Will You place upon it one who causes corruption therein and sheds blood, while we exalt You with praise and declare Your perfection?"*

This verse serves as a testament to the angels' observation of the Bashar or pre-humans on Earth engaging in acts of bloodshed and violence. This observation aligns with contemporary scientific findings, especially in the fields of archaeology and paleontology, which offer substantial evidence of such behaviors among early hominins. The archaeological and fossil records provide traces of violence, including indicators of interpersonal conflicts, territorial disputes, and even instances of cannibalism.

Adam – The First Human being

The story of Adam in the Quran, primarily a theological narrative, doesn't readily align with scientific understanding. Nevertheless, certain misconceptions surrounding the story have led many Muslim scholars and followers to reject the concept of human evolution. Through meticulous research, answers within the Quran have been unearthed, allowing for the reconstruction of the story of Adam in accordance with divine commentary.

Adam's Parents: The Quran does not explicitly state that Adam was born without parents. The verse often cited to suggest Adam's creation without parents is Chapter 3 Verse 59, which compares the creation of Jesus and Adam. It states, *"Indeed, the example of Jesus to God is like that of Adam. He created Him from dust; then He said to him, "Be," and he was."* This implies that both Adam and Jesus were created through both types of divine creation, Alam-e-Khalq (the World of Creation) and Alam-e-Amr (the World of Command), rather than asserting that Adam had no parents and Jesus had no father.

According to this perspective, Adam and Eve, like the other people accompanying them, had parents. This interpretation aligns with the understanding that the verse is emphasizing the divine creation of both Jesus and Adam, rather than the absence of parental lineage for either figure.

Adam as Father of Humanity: In the Quran, human beings are referred to as the children of Adam, known as Bani Adam in Arabic. This designation doesn't imply that all human beings are direct biological descendants of Adam and Eve. Instead, it aligns with a cultural and tribal naming tradition prevalent in Arab societies. The custom involves naming tribes or clans after influential or significant figures within the community.

An analogous practice is seen in the time of Prophet Yaqoob (Israel). People who accepted religion on his invitation, even if not biologically related to Bani Israel, became integrated into the tribe over time. Similarly, the use of 'Bani Adam' in the Quran aims to establish a collective identity for all humankind, associating them symbolically with the very first human being, Adam. This linkage emphasizes that humanity shares a collective identity with Adam, regardless of not being direct biological descendants.

The Quranic narrative on Human Evolution

Adam Created on Earth: The research highlights and corrects another misconception regarding the place of creation of Adam. According to the Quran, Adam was created directly on Earth, as God informed the angels about the creation of Bashar and the subsequent creation of human beings, appointing them as Khalifa on Earth. There is no deviation from this narrative in any Quranic verse.

The misinterpretation that led to the concept of Adam being created in paradise, undergoing a test with Eve in paradise, and facing punishment by being expelled from paradise to Earth is addressed and clarified through the research. The Quranic verses do not support the notion of Adam's initial creation in paradise and subsequent descent to Earth as a punishment.

Chapter 7 Verse 19 states, *"And [O Adam], dwell, you and your wife, in the <u>Jannah</u> and eat from wherever you will but do not approach this tree, lest you be among the wrongdoers."* It's important to note that the term "Jannah" - (الجَنَّة) is often mistranslated as "paradise." However, this word is used more than 15 times in the Quran to refer to an earthly garden rather than the celestial paradise.

Adam and Eve were not dwelling in the ultimate paradise but in an earthly garden on Earth. This garden provided them with an abundance of food, water, and shade. It's crucial to distinguish this earthly garden from the paradise promised for the righteous after the Day of Judgment, where believers will stay eternally without the risk of expulsion. The Quran, in Chapter 3 Verse 198, states, *"But those who feared their Lord will have gardens (جَنَّٰتٌ) beneath which rivers flow, abiding <u>eternally</u> therein, as accommodation from God. And that which is with God is best for the righteous."*

Chapter 7 Verse 24 includes, *"<u>Go down</u>, being enemies to one another,"* and Chapter 20 Verse 123 of the Quran encompasses, *"He (God) said, <u>Go down</u> from it (the garden)."* The terms 'Ahbitu' (اهْبِطُوا) and 'Ahbita' (اهْبِطَا) are translated as "Go down" or "Descend" in these verses. However, the same word is used in another verse of the Quran. Chapter 2 Verse 61 recounts the interaction between Moses and the

Bani Israel, *"He (Moses) said, What! You wish to exchange the better for something inferior? Therefore, settle down (اهْبِطُوا) in Egypt or any city, where you will get what you demand."*

Keeping the later verse in mind, Adam and Eve were instructed to leave the garden and settle down elsewhere on Earth.

Adam and Other Cohabitants: The Quran underscores on two occasions that Adam was selected from a broader population of Bashar as a representative of his kind. Initially, he was chosen to be the first human being, and later he was appointed as a prophet of God. Chapter 3 Verse 33 states, *"Indeed, God chose Adam and Noah and the family of Abraham and the family of Imran over the worlds"* and Chapter 20 Verse 122 states, *"Then his Lord chose him, and turned towards him with forgiveness, and granted him guidance."*

Moreover, Quran explicitly addresses people in the plural form of grammar in Chapter 20 Verse 123, when Adam and Eve were instructed to leave the garden, *"He (God) said, Go down from it, <u>all of you</u>. And when guidance comes to you from Me, whoever follows My guidance, there will be no fear concerning them, nor will they grieve"* and Chapter 7 Verse 24, same situation, people are being addressed, *"Go, settle down, <u>some of you enemies of some.</u>"*

The verses unequivocally suggest that Adam and Eve were not solitary inhabitants in the garden. The first verse, emphasizing success for those who follow guidance, inherently targets human beings rather than Satan, who had already been cast out. Similarly, the second verse, frequently associated with Satan, employs the term "some of you to some" (بَعْضُكُمْ لِبَعْضٍ), challenging the notion of Satan's involvement, as it would necessitate multiple Satans to qualify as "some". Furthermore, the proclamation that "some of you will be enemies of some" directly alludes to human beings, as Satan had already been declared humanity's ultimate adversary.

PART III: Conclusion

Unbiased Pursuit of Knowledge

In the relentless quest for knowledge, maintaining an unbiased approach becomes the linchpin for a thorough and meaningful exploration. Whether navigating the narratives in the Quran or delving into scientific perspectives on human evolution, an open mind devoid of preconceived notions is paramount.

In the multifaceted realm of human knowledge, the call for an unbiased pursuit is akin to embarking on a grand intellectual odyssey. The canvas upon which the narratives of the Quran and the tenets of scientific understanding unfold is vast and intricate. The pursuit begins with a conscious decision to lay aside preconceived notions, recognizing that the journey into understanding is best undertaken with the clarity of an unencumbered mind.

Unbiased pursuit doesn't imply a detachment from personal beliefs; rather, it advocates for a respectful coexistence of faith and reason. It encourages the seeker to engage in a dance between reverence for spiritual insights and an appreciation for the empirical methodologies of science. To navigate these waters successfully, one must be willing to set sail without predetermined destinations, allowing the currents of evidence and revelation to guide the ship of inquiry.

Illuminating Quranic Revelations

The revelation of the concept of evolution within the Quran, illuminated through careful research, promises to chart novel pathways for exploration, seamlessly intertwining scientific and theological dimensions. What renders this research truly captivating is that the Quran, revealed fourteen centuries ago, articulates evolutionary concepts with remarkable clarity and confidence, akin to the narrative of an eyewitness. The unique perspective of the Quran, devoid of speculation and presented as definitive occurrences aligned with divine will, gives rise to the concept aptly termed "Divine Evolution of Human being."

In delving into the luminous realm of the Quranic revelations, we find a narrative that transcends the temporal constraints of its revelation. The Quran's insights into the concept of evolution radiate a timeless brilliance, inviting contemplation across the ages. The verses weave a narrative that resonates not just with the era of their revelation but spans the epochs of human understanding, echoing the profound truths of existence.

The clarity and confidence with which the Quran articulates the concept of evolution stand as a testament to the divine wisdom embedded in its verses. It is as if the Quran, in unveiling the intricacies of life's evolution, extends an invitation to humanity to explore the wonders of creation with open hearts and discerning minds. The revelations, presented as a tapestry of divine guidance, beckon believers to marvel at the unfolding drama of existence and recognize the hand of God orchestrating the symphony of life.

Especially poignant is the Quran's portrayal of evolution as a manifestation of divine will. The verses do not merely describe a sequence of events; they assert the intentional unfolding of creation as per the divine plan. This portrayal transforms the concept of evolution from a series of random occurrences to a purposeful and guided journey—a testament to the unfolding wisdom of God.

The Quranic narrative surpasses the confines of creationism, painting a picture of evolution as a complex and intricate manifestation of divine design. The instantaneous creation gives way to a process guided by divine orchestration. The portrayal serves as an invitation to contemplate the profound nature of divine evolution—a journey where each stage unfolds with purpose, intricately interwoven into the grand narrative of creation. The Quranic perspective challenges believers to transcend simplistic notions and embrace the divine complexity inherent in the ongoing evolution of life.

The Constant of the "Will of God"

In the delicate interplay between scientific exploration and theological contemplation, the introduction of the constant of the "Will of God" offers a profound avenue for synthesis. This constant serves as a bridge, acknowledging the coexistence of scientific theories and spiritual insights. It recognizes that the intricacies observed by science are not random happenings but rather guided by divine intentionality. This conceptual anchor becomes a unifying force, connecting the empirical observations of science to the overarching purpose set forth by divine will.

The harmonization facilitated by the constant encourages a holistic understanding, bridging the perceived gap between science and theology. It transforms discussions on human evolution from a dichotomy of belief or disbelief into a richer dialogue where empirical observations and spiritual insights complement each other. This approach invites believers and seekers to contemplate the wonders of the natural world without compromising their spiritual convictions, fostering an environment where both realms contribute to a more profound reality. Incorporating the element of Divine Will into any scientific theory can render it more acceptable, or, at the very least, elicit a more receptive response from believers. For instance, suggesting that Human Evolution occurred in alignment with "Will of God", or proposing that apes are the ancestors of humankind as per "Will of God", allows for a harmonious integration of spiritual beliefs with scientific concepts.

The constant of the "Will of God" transcends dichotomous thinking, encouraging a collaborative exploration that seeks wisdom from diverse sources. By substituting terms like "divine selection" for "natural selection" and "mutation by divine will" for "random mutation," the constant aligns the language of science with theological frameworks, nurturing a nuanced exploration. It fosters a shared pursuit of truth, steering both scientific and theological inquiries away from dogmatism and toward a collective journey of discovery.

Science and Interpretation of the Quran

The research presented in this book serves as a refined distillation of insights derived from the Quran. Yet, understanding its profound significance requires more than a straightforward approach—it demands the underpinning of empirical-based scientific theories and established facts. This echoes the broader trend where various acknowledgements of scientific facts within the Quran were initially construed metaphorically. From a theological standpoint, science is perceived as a creation of God, leading to the assertion that any scientific discovery aligning with holy scripture is often presented as a claim affirming the truth of the scripture itself.

A frequently posed question arises when a novel interpretation of religious scripture is introduced: why was it not revealed earlier? Often, the reluctance to accept a new interpretation stems from the absence of endorsement by well-renowned scholars, both in the present and the past. This skepticism can be likened to someone expressing disbelief in the concept of gravitational force as presented by Isaac Newton simply because no one in the past articulated it. However, the denial of a concept, be it gravitational force or interpretations of religious scriptures, does not negate its existence. Just as the force of gravity has perpetually existed, the understanding embedded in religious scriptures might only be revealed at a particular moment, yet the essence of the concept has always been inherent.

Scientific discoveries have wielded a profound impact on the comprehension of numerous verses in the Quran. The examples elucidated under the section "Quran and Science" in Chapter One of this book resonate with revelations brought forth by scientific advancements. Preceding these discoveries, interpretations of these verses may have taken on a different hue, perhaps exclusively from a religious perspective or in a metaphorical context. The stages of zygote development, intricately embedded in the Quran for over 14 centuries, remained veiled until recently, when Dr. Maurice Bucaille, an eminent French medical doctor, brought them to light. It was only through the lens of contemporary scientific revelations that these profound insights within the Quran were corroborated and unearthed, underscoring the dynamic interplay between religious scripture and scientific understanding.

Limitations of Science

While science has made remarkable strides in unravelling the intricacies of evolution, it is essential to acknowledge the inherent limitations within the scientific framework when exploring this complex phenomenon. One significant constraint lies in the inability to directly observe and document the entirety of evolutionary processes, particularly those spanning millions of years. The vast timescales involved in evolution often exceed the temporal scope of human observation and documentation, leaving scientists to rely on fragmentary evidence and indirect inference.

Furthermore, the incomplete fossil record poses a substantial hurdle in constructing a comprehensive narrative of evolution. Fossilization is a rare occurrence, and the preservation of an organism in the geological record is contingent on specific environmental conditions. Consequently, many transitional forms and intermediate species that could provide crucial insights into evolutionary pathways may never be discovered due to the limitations of fossilization. The fossil record, though invaluable, represents a mere snapshot of the evolutionary timeline, leaving significant gaps that challenge a thorough understanding of the intricate web of life's development.

Another limitation arises from the dynamic and contingent nature of evolutionary processes. Evolution is influenced by multifaceted interactions between genetics, environment, and chance events, making it challenging to create precise predictive models. While scientific theories offer compelling explanations for observed phenomena, the inherent complexity of biological systems introduces uncertainties that cannot be entirely eliminated. As a result, the predictive power of evolutionary models is constrained by the intricate interplay of factors, underscoring the continuous refinement and adaptation of scientific understanding in the quest to unravel the mysteries of evolution.

Impact of new Scientific discoveries

New discoveries in science have the transformative power to challenge and reshape prevailing theories related to evolution. One fundamental aspect is the dynamic nature of scientific knowledge, where new findings can provide fresh perspectives, introduce novel information, and even necessitate a re-evaluation of existing paradigms. These discoveries can lead to a shift in our understanding of evolution, either refining current theories or, in some cases, rendering them obsolete.

Advancements in technologies, particularly in fields like genetics and paleontology, have the potential to unearth previously inaccessible information. For instance, the discovery of new fossil evidence or the development of advanced genetic sequencing techniques can shed light on previously unknown aspects of evolutionary history. These breakthroughs may reveal new branches in the tree of life, identify transitional forms, or provide insights into the mechanisms driving evolutionary processes.

Furthermore, interdisciplinary collaborations and the integration of diverse scientific disciplines can open new avenues for exploration. Discoveries in fields such as ecology, climatology, or microbiology may offer insights that challenge or expand existing evolutionary models. The interconnectedness of various scientific domains allows for a more holistic understanding of the factors influencing evolution.

In some instances, new discoveries may challenge prevailing theories by revealing exceptions or anomalies that were not previously accounted for. These anomalies can prompt scientists to reconsider their assumptions and refine their models to accommodate the newfound complexities. The scientific process encourages a dynamic dialogue where theories are continuously tested, refined, or even replaced as our knowledge evolves.

In essence, the fluidity of scientific discovery ensures that our understanding of evolution remains open to revision. While established theories provide a framework for interpretation, the ever-expanding realm of scientific exploration constantly invites the possibility of paradigm shifts, where new discoveries have the potential to reshape our comprehension of the intricate tapestry of life's evolutionary history.

Lack of Science Education

In a historical context, Muslim scholars were pioneers in the realms of invention and scientific discovery, with Muslim states hosting some of the world's leading universities. These institutions attracted students globally, fostering an environment of modern education. However, over time, there has been a noticeable decline in this trajectory. The contemporary educational landscape, particularly the Madrasa System responsible for cultivating Muslim scholars, is grappling with deficiencies, notably in the provision of comprehensive science education. Regrettably, the system not only falls short in imparting up-to-date knowledge but also tends to discourage the application of scientific logic in understanding natural phenomena. Consequently, emerging Muslim scholars often find themselves inadequately equipped with the requisite knowledge for navigating the complexities of the modern world.

As a result of the prevailing limitations within the Madrasa system, scholars often find themselves ill-equipped to address inquiries related to modern science. The scholars who manage to make a substantial impact are frequently those who embark on a personal journey of learning religion after acquiring conventional education, distinct from the Madrasa system. Notable figures such as Molana Modudi, a source of inspiration for many, did not emerge from the Madrasa system. Dr Asrar Ahmed, who continues to captivate the youth with his insights into modern questions, pursued a career in medicine. Muhammad Ali Mirza, an engineer by profession, serves as a valuable resource for contemporary knowledge seekers. Dr Zakir Naik, a medical doctor by profession, generally encourages scientific interpretation of the Quran. These examples underscore the limited presence of Muslim scholars within the Madrasa system who actively engage with scientific knowledge. Dr Tahir ul Qadri, with a Ph.D. degree, stands as a rare exception within this context.

A sincere seeker of knowledge often poses genuine questions and, at times, finds themselves grappling with increased perplexity. While not everyone embarks on a personal journey to seek answers independently, the absence of scientific knowledge among scholars can potentially lead questioners astray or leave them with unaddressed doubts. The imperative lies in furnishing emerging religious scholars with a comprehensive, modern education that not only benefits their personal understanding but also empowers them to guide others effectively. This shift towards a broader education aims to bridge the gap between traditional religious teachings and the complexities of the contemporary world. By equipping scholars with a diverse range of knowledge, including the sciences, there emerges a more informed and capable cadre of religious leaders capable of navigating the intricacies of both religious doctrine and the advancements of the modern era.

Blind Faith and Polarization

Regrettably, the acceptance or rejection of the concept of Human Evolution has metamorphosed into a matter of blind faith. Instead of fostering an open and inclusive dialogue, the subject has morphed into a divisive tool, often wielded to hastily categorize individuals as either atheists or illiterate. The rich tapestry of nuanced complexities that should define discussions on human evolution is overshadowed by polarizing labels, creating an environment that obstructs meaningful conversations and impedes the exploration of diverse perspectives on this intricate aspect of our understanding of existence.

In the broader discourse surrounding human evolution, the polarization of views has cast a shadow over the nuanced exploration that such a complex topic demand. The dichotomy between acceptance and rejection becomes a contentious battleground where individuals are swiftly labelled, thus undermining the richness of thought within both religious and scientific communities. This oversimplification of the complex interplay between faith and reason stifles the potential for mutual understanding and collaboration in unravelling the mysteries of human existence. By reducing the discourse to binary oppositions, the true depth and diversity of perspectives that could contribute to a more holistic understanding of human evolution are often overlooked, perpetuating a cycle of polarization that hinders the genuine exchange of ideas and insights.

In navigating the intricate landscape of human evolution, it becomes imperative to transcend the limitations of entrenched positions and foster an environment where divergent perspectives can coexist and contribute to a more comprehensive understanding. Breaking free from the constraints of blind faith and polarized viewpoints allows for a more inclusive and enriched exploration of the complexities inherent in the evolution of humanity.

Persecution and Adversity

A disconcerting reality unfolds globally as individuals, particularly educators, find themselves ensnared in persecution based on their embrace or rejection of the concept of human evolution. In certain regions, educators who align with the tenets of human evolution face adversity and hostility, often impeding their ability to engage in open and constructive dialogue with students. Conversely, in different corners of the world, individuals advocating alternative concepts such as Intelligent Design or creationism encounter their own share of adversity, as divergent perspectives are met with resistance and skepticism.

The challenges faced by educators and individuals advocating diverse perspectives underscore the broader issue of intellectual freedom and the essential need for a diversity of thought in shaping our understanding of the world. In certain regions, the pursuit of knowledge is met with hostility, hindering the free exchange of ideas and impeding the intellectual progress of societies. Academic environments that discourage the exploration of alternative concepts stifle the intellectual curiosity that drives discovery and innovation. Conversely, in other areas, the advocacy of alternative concepts encounters adversity, limiting the intellectual diversity necessary for a robust and evolving understanding of the world.

As the global discourse on the acceptance or rejection of concepts such as human evolution unfolds, it becomes evident that the challenges faced by educators and proponents of diverse perspectives are not isolated incidents but indicative of broader issues surrounding intellectual freedom. Navigating these challenges requires a concerted effort to promote environments where diverse viewpoints can coexist, fostering an atmosphere that encourages the exploration of ideas and the pursuit of knowledge without fear of persecution or marginalization.

Healthy Debates

The intellectual debates between proponents and opponents of the concept of Evolution often culminate in both sides claiming victory, leaving viewers, with genuine intentions to learn, without substantial knowledge gained. The debates, intended to provide a platform for discussions, often transform into battlegrounds for ideological clashes rather than fertile grounds for knowledge expansion. The intricacies of scientific concepts and theological insights become casualties in the quest for ideological vindication, hindering the true pursuit of understanding.

However, engaging in healthy debates may provide a platform for the exchange of diverse perspectives, fostering a more comprehensive understanding of the intricacies surrounding the concept. Scientists bring empirical evidence and data-driven insights to the table, drawing on observations, experiments, and the scientific method to support their claims. On the other hand, theologians contribute a philosophical and theological framework, exploring questions of purpose, meaning, and divine intervention in the evolutionary process.

The clash of ideas in these debates should serve as a crucible for refining and enhancing both scientific and theological understandings. It may prompt scientists to critically evaluate their theories, ensuring they withstand rigorous scrutiny and align with observed phenomena. Similarly, theologians may be challenged to reconcile their beliefs with the empirical evidence presented by scientists, fostering a nuanced interpretation of religious texts. Ultimately, healthy debates can cultivate an environment where both disciplines can coexist, each contributing unique insights without compromising the integrity of the other.

Moreover, these debates have the potential to bridge gaps between seemingly disparate realms of knowledge. By encouraging dialogue, scientists and theologians can find common ground and shared understanding, fostering collaboration rather than conflict. This synergy promotes a more holistic approach to addressing questions about the origin and development of life, acknowledging that science and theology, when harmonized, can enrich our collective exploration of the complex tapestry of evolution.

About the Author

Saeed Rajput, an accomplished corporate banker holding prestigious degrees in MBA and Master of Finance, currently resides in the UK, originally hailing from Pakistan. With a profound dedication to the study of the Quran, Saeed considers himself a devoted student of its teachings, accumulating 15 years of in-depth exploration. His scholarly journey led him to five years of intensive research, culminating in the creation of his latest work, this book.

Alongside his academic pursuits, Saeed has explored the realm of poetry, showcasing his creative expression in Urdu through the publication of two poetry books.

Exploring the genetic intricacies that define his identity, Saeed's paternal lineage reveals a Y chromosomal haplogroup of R-F992, and on the maternal side, his DNA showcases an MT DNA haplogroup of F1. Remarkably, Saeed's genetic makeup includes 2.43% Mezmaiskaya 1 Neanderthal DNA and 0.18% Denisovan DNA, offering intriguing insights into his ancestral heritage and evolutionary lineage.

Unraveling the tapestry of his ancestry, the last 1500 years connect Saeed to the Indian Subcontinent, with roots deeply embedded in Pashtun and Sindhi heritage. A unique thread in his maternal lineage traces back to the lost tribe of Israel, specifically to the descendants of Manasseh, son of Joseph. On his paternal side, Saeed's lineage can be traced back to the esteemed Raja Dahir of Sindh, renowned for his principled stance against Mohammad bin Qasim during the Umayyad dynasty's era. This historical context revolves around Raja Dahir's refusal to return the Ehl-e-Bait, the family of Prophet Muhammad, seeking sanctuary from Umayyad oppression.

Saeed takes pride in sharing his diverse journey, weaving together strands of academic rigor, poetic expression, and a profound connection to his ancestral heritage. Through his works, he aspires to contribute to the intersection of knowledge, culture, and spirituality.

Bibliography

Lamarck, J.-B. (1907). Philosophie Zoological Philosophy: An Exposition with Regard to the Natural History of Animals. Macmillan and Co.

Darwin, C. (1859). *On the Origin of Species by Means of Natural Selection*. n.d.

Mayr, E. (1982). *The Growth of Biological Thought: Diversity, Evolution, and Inheritance*. Harvard University Press.

Browne, J. (1995). *Charles Darwin: Voyaging*. Princeton University Press.

Darwin, F. (Ed.). (1887). *The Life and Letters of Charles Darwin*. John Murray.

Futuyma, D. J. (2017). *Evolution*. Sinauer Associates.

Freeman, S., & Herron, J. C. (2019). *Evolutionary Analysis*. Pearson.

Gould, S. J., & Eldredge, N. (1977). *Punctuated equilibria: The tempo and mode of evolution reconsidered*. Paleobiology, 3(2), 115-151.

Eldredge, N., & Gould, S. J. (1997). *On punctuated equilibria*. In S. M. Stanley (Ed.), *Macroevolution: Pattern and Process* (pp. 82-115). The University of Chicago Press.

Stanley, S. M. (1979). *Macroevolution: Pattern and Process*. W. H. Freeman.

Numbers, R. L. (2006). *The Creationists: From Scientific Creationism to Intelligent Design*. Harvard University Press.

Morris, H. M. (2007). *The Genesis Record: A Scientific and Devotional Commentary on the Book of Beginnings*. Baker Books.

Dembski, W. A., & Ruse, M. (2004). *Debating Design: From Darwin to DNA*. Cambridge University Press.

Behe, M. J. (1996). *Darwin's Black Box: The Biochemical Challenge to Evolution*. Free Press.

Johnson, P. E. (1991). *Darwin on Trial*. InterVarsity Press.

Forrest, B., & Gross, P. R. (2004). *Creationism's Trojan Horse: The Wedge of Intelligent Design*. Oxford University Press.

Alberts, B., Johnson, A., Lewis, J., Raff, M., Roberts, K., & Walter, P. (2014). *Molecular Biology of the Cell*. Garland Science.

Ridley, M. (2006). *Genome: The Autobiography of a Species in 23 Chapters*. Harper Perennial.

Stoneking, M., & Krause, J. (2011). *Genetics and human evolution*. In A. M. Weiss, & J. E. Buikstra (Eds.), *Biosocial Criminology: New Directions in Theory and Research* (pp. 245-269). Routledge.

Ridley, M. (2004). *Evolution*. Blackwell Publishing.

Mayr, E. (2001). *What Evolution Is?* Basic Books.

Al-Jahiz. (N.d.). *Kitab al-Hayawan (The Book of Animals)*. n.d.

Lunde, P. (Translator). (1969). *The Book of Animals: Al-Jahiz - A Parallel Arabic-English Text*. Volume 3. University of California Press.

Leaman, O., & Nasr, S. H. (2001). *History of Islamic Philosophy*. Routledge.

Miskawayh, A. A. (N.d.). *Tahdhib al-Akhlaq (The Refinement of Character)*. n.d.

Leaman, O. (2000). *A Brief Introduction to Islamic Philosophy*. Polity.

Fakhry, M. (2004). *A History of Islamic Philosophy*. Columbia University Press.

Ibn Khaldun. (n.d.). *Al-Muqaddimah (The Introduction)*. n.d.

Rosenthal, F. (1958). *Ibn Khaldun: The Muqaddimah: An Introduction to History*. Princeton University Press.

Dhaouadi, M. (2013). "Ibn Khaldun." *Stanford Encyclopedia of Philosophy*. Retrieved from https://plato.stanford.edu/entries/arabic-islamic-language/

Dawkins, R. (1986). *The Blind Watchmaker: Why the Evidence of Evolution Reveals a Universe Without Design*. W. W. Norton & Company.

Sahih International. (n.d.). The Noble Quran. Sahih International.

Abul Ala Maududi (n.d.). *Tafheem-ul-Quran (English translation and commentary)*. Islamic Publications

Dr. Muhammad Tahir-ul-Qadri (n.d.). *Irfan-ul-Quran*. Minhaj-ul-Quran Publications

Dr. Asrar Ahmed (n.d.). *Explanation of Quran in Urdu*. Retrieved from https://www.youtube.com/watch?v=2WAFIAfL7nM&list=PLLSbPBshw4tBeAXprgSWX8g-7PWmG28n4

Maulana Syed Mumtaz Ali (2008). Esharia Mazameen-e-Quran. Al-Faisal Nashran. ISBN: 969-503-084-X

Wikipedia contributors. (2023, January 10). *Cytoplasm*. In Wikipedia. https://en.wikipedia.org/wiki/Cytoplasm

Wikipedia contributors. (2023, January 15). *Big Bang*. In Wikipedia. https://en.wikipedia.org/wiki/Big_Bang

Islamic Awareness. (n.d.). *Scientist's Comments on the Quran*. Islamic Awareness. http://www.islamic-awareness.org/quran/science/scientists.html

Walrath, M. E. (n.d.). *History of the Earth's Formation*. ISBN-13: 978-0530178776

Davis, R.A. (n.d.). *Principles of Oceanography*. ISBN-13: 978-0201014648

Wikipedia contributors. (2023, December 18). *Orbit*. In Wikipedia. https://en.wikipedia.org/wiki/Orbit

Wikipedia contributors. (2023, November 13). *Celestial mechanics*. In Wikipedia. https://en.wikipedia.org/wiki/Celestial_mechanics

Press, F. (n.d.). Earth. ISBN-13: 978-071671743.

Wikipedia contributors. (2023, December 14). *Expansion of the universe*. In Wikipedia. https://en.wikipedia.org/wiki/Expansion_of_the_universe

Shipton, E. A. (n.d.). *Skin Matters: Identifying Pain Mechanisms and Predicting Treatment Outcomes*. Neurol Res Int., 2013, 329364.

Wikipedia contributors. (2023, December 20). *Fingerprint*. In Wikipedia. https://en.wikipedia.org/wiki/Fingerprint

Mendelian inheritance. (n.d.). *Mendelian inheritance*. In Encyclopedia Britannica. https://www.britannica.com/science/Mendelian-inheritance

Wikipedia contributors. (2023, April 18). *Neo-Darwinism*. In Wikipedia. https://en.wikipedia.org/wiki/Neo-Darwinism

New World Encyclopedia. (n.d.). *Stephen Jay Gould*. In New World Encyclopedia. https://www.newworldencyclopedia.org/entry/Stephen_Jay_Gould

OpenMind. (2017, March 16). *Dark Matter: The Mystery Substance Physics Still Can't Identify* [Video]. YouTube. https://www.youtube.com/watch?v=HpXaiG8L28s

Wikipedia contributors. (2023, December 20). *Human taxonomy*. In Wikipedia. https://en.wikipedia.org/wiki/Human_taxonomy

Encyclopedia Britannica. (2023, December 23). *Background and beginnings in the Miocene*. Britannica.

https://www.britannica.com/science/human-evolution/Background-and-beginnings-in-the-Miocene

A-Z Animals. (n.d.). *The 10 Oldest Human Fossils Ever Found. A-Z Animals.* https://a-z-animals.com/blog/the-10-oldest-human-fossils-ever-found/

Miller, S. L., & Urey, H. C. (1959). *Organic Compound Synthesis on the Primitive Earth.* Science, 130(3370), 245-251. DOI: 10.1126/science.130.3370.245

Shklovskii, I. S., & Sagan, C. (1966). *Intelligent Life in the Universe.* Holden-Day.

Wikipedia contributors. (2023, December 17). *Last universal common ancestor.* In Wikipedia. https://en.wikipedia.org/wiki/Last_universal_common_ancestor

Walker, R. S., & Shipman, P. (2015). *Interpersonal Violence in the Late Pleistocene.* Nature.

Garcia-Romeu, J. A., & Konigsberg, L. W. (2017). *A 100-Year Review of Cannibalism in Modern Literature: What Are the Odds? American* Journal of Physical Anthropology.

Arkush, E. M. (2008). *The Archaeology of Warfare: Prehistories of Raiding and Conquest.* University Press of Florida.

NHM. (2023, October). *Oldest evidence of human cannibalism as a funerary practice.* Natural History Museum, https://www.nhm.ac.uk/discover/news/2023/october/oldest-evidence-of-human-cannibalism-as-a-funerary-practice.html

Alusi, M. S. A. (n.d.). *Ruh al-Ma'ani (Translated by: Sayyid Shah Rafi al-Din).* n.d.

National Geographic. (n.d.). *Scientific Adam and Eve | National Geographic* [Video]. YouTube. https://www.youtube.com/watch?v=G3LLtZUdpvA

www.ingramcontent.com/pod-product-compliance
Lightning Source LLC
Chambersburg PA
CBHW030438010526
44118CB00011B/685